Nanofabrication: Principles to Laboratory Practice

Nanofabrication: Principles to Laboratory Practice

Editor

Mudiwa Afolayan

scitus
academics

Nanofabrication: Principles to Laboratory Practice
Edited by **Mudiwa Afolayan**

ISBN: 978-1-68117-718-2
Library of Congress Control Number: 2016958822

SCITUS Academics LLC,
616, Corporate Way, Suite 2, 4766,
Valley Cottage, NY 10989

www.scitusacademics.com

Notice

Printed in United States of America on Acid Free Paper

Preface

Nanofabrication is the design and production of nanometer measured devices. Nanofabrication technology defines the technology's leading edge. Nanofabrication is of interest to computer engineers as the door is opened to microprocessors and memory chips with super-high density. However, there are a number of ways nanofabrication can be done. Nanofabrication is the technology that has emerged from the manufacture of chips for semiconductor microelectronics. It is used today, of course, in information storage and in bio-medical applications, pharmaceuticals, power semiconductors, and micro-electro-mechanical equipment, sensors, opto-electronics, and microelectronics. It is the driver technology in areas such as informatics, the Internet, computers and bio-medical technology.

This develops a fundamental understanding of the processes and applications of commonly used device manufacturing in the R&D environment. Nanodevices and Micro-have many advantages over their complements of macroscale. For applications in the semiconductor industry, microfabrication techniques have been developed and are therefore not specific to medical or biological applications. Nevertheless, micro-and nano-manufacturing have offered a number of options for the study of chemical, biological and physical processes on a molecular and cell scale and also for the style of synthetic devices that interact effectively with natural systems at these levels. This book is intended for the practical scientist or engineer, with the aim of transferring the information from the book to the laboratory.

Table of Contents

1

CHAPTER

Maskless Micro/Nanofabrication on Gaas Surface by Friction-Induced Selective Etching

Peng Tang, Bingjun Yu, Jian Guo, Chenfei Song and Linmao Qian

Tribology Research Institute, National Traction Power Laboratory, Southwest Jiaotong University

ABSTRACT

In the present study, a friction-induced selective etching method was developed to produce nanostructures on GaAs surface. Without any resist mask, the nanofabrication can be achieved by scratching and post-etching in sulfuric acid solution. The effects of the applied normal load and etching period on the formation of the nanostructure were studied. Results showed that the height of the nanostructure increased with the normal load or the etching period. XPS and Raman detection demonstrated that residual compressive stress and lattice densification were probably the main reason for selective etching, which eventually led to the protrusive nanostructures from the scratched area on the GaAs surface. Through a homemade multi-probe instrument, the capability of this fabrication method was demonstrated by producing various nanostructures on the GaAs surface, such as linear array, intersecting parallel, surface mesas, and special letters. In summary, the proposed method provided a straightforward and more maneuverable micro/nanofabrication method on the GaAs surface.

Keywords: Maskless Micro/nanofabrication GaAs Friction-induced selective etching

BACKGROUND

Due to its direct bandgap and high electron mobility, gallium arsenide (GaAs) has become one of the most widely used compound semiconductor materials. For instance, GaAs is the perfect substrate for quantum luminescent devices, such as photoelectric detector [1], high-performance laser [2], quantum information processing [3], and so on. Nevertheless, the precondition for realizing these quantum devices is to grow quantum dots on certain positions of substrate [4, 5]. Thus, the controllable fabrication of the patterned GaAs substrate is a significant issue of concern.

Many efforts have been made in producing patterned GaAs substrate. As a prevalent technology, the photolithography has been used for the fabrication on GaAs surface [6]. However, the conventional photolithography has the embarrassment in its resolution vs. cost, i.e., the production cost will be much higher as the resolution is improved [7]. Although the electron beam (EB) [8] and focused ion beam (FIB) [9] lithography technology can enable higher machining precision and finer resolution patterning, these techniques are costly and complex, requiring multiple-step processes [10]. In addition, anodic oxidation has been demonstrated the potential in the creation of structures, patterns, and devices at the nanometer scale [11], but the operation will be controlled under rigorous environmental requirements, such as applied voltage, suitable humidity, and so on [12]. Such processes can also bring contamination and impurity onto the area fabricated [13]. In recent decades, the proximal probe method based on the mechanical stamp and scratching technique has been employed to produce patterned GaAs substrate [4, 14], but it is difficult, if not impossible, to fabricate GaAs nanostructures with low destruction by solely mechanical scratching. Therefore, it is necessary to develop a straightforward and more flexible fabrication method for the GaAs surface.

In the present study, a novel friction-induced micro/nanofabrication method that consists of nanoscratching and post-etching was presented to produce nanostructures on GaAs. The effects of the applied normal load and etching period on the formation of the nanostructure were studied. Based on the X-ray photoelectron spectroscope (XPS) and Raman spectra characterization, the fabrication mechanism of the nanostructure was discussed. Finally, through a homemade multi-probe instrument, the capability of this fabrication method was demonstrated by producing various nanostructures on the GaAs surface, such as linear array, intersecting parallel, surface mesas, and special letters.

METHODS

Material

The GaAs (100) wafers, n-doped with Si, were purchased from JMEM Electronic Materials, Ltd., Tianjin, China. Using an atomic force microscope (AFM, SPI3800N, Seiko, Tokyo, Japan), the surface root-mean-square (RMS) roughness of the GaAs wafer was measured as 0.5 nm over a 1 μm × 1 μm area. The crystal state of the GaAs material was detected by the X-ray diffraction (XRD, X'Pert, PANalytical, Almelo, Netherlands), showing that the GaAs wafer was single crystal in (100) plane orientation. Before the fabrication, the GaAs wafers were ultrasonically cleaned with methanol and ethanol for 3 min in turn, and successively rinsed with deionized water for 10 min to remove surface contamination.

Fabrication method

As shown in Figure 1, the maskless fabrication process consists of scratching and post-etching. When the GaAs surface was scratched by a diamond tip along the designed traces, grooves can be generated on the scanned area. After etching in H_2SO_4 aqueous solution, different protrusive nanostructures can be produced *in situ* from the scratched area on the GaAs surface. Scratching tests on the GaAs surface were performed by a nanoscratch tester (CSM Instruments, Peseux, Switzerland) or a homemade multi-probe instrument [15]. The spherical diamond tips used for scratching have the radii of about 5 μm. After the scratching tests, the specimens were dipped in a mixture of H_2SO_4 aqueous solution ($H_2SO_4/H_2O_2/H_2O = 1:0.5:100$) for post-etching [16]. During scratching and post-etching, the experimental temperature was controlled at 22°C and the relative humidity varied between 50% and 55%. All the AFM images of GaAs specimens were scanned by silicon nitride tips (MLCT, Veeco Instruments Inc., Plainview, NY, USA) with the spring constant $k = 0.1$ N/m in vacuum. The morphology of GaAs surface patterns was observed by a scanning electron microscope (SEM, QUANTA200, FEI, Hillsboro, OR, USA).

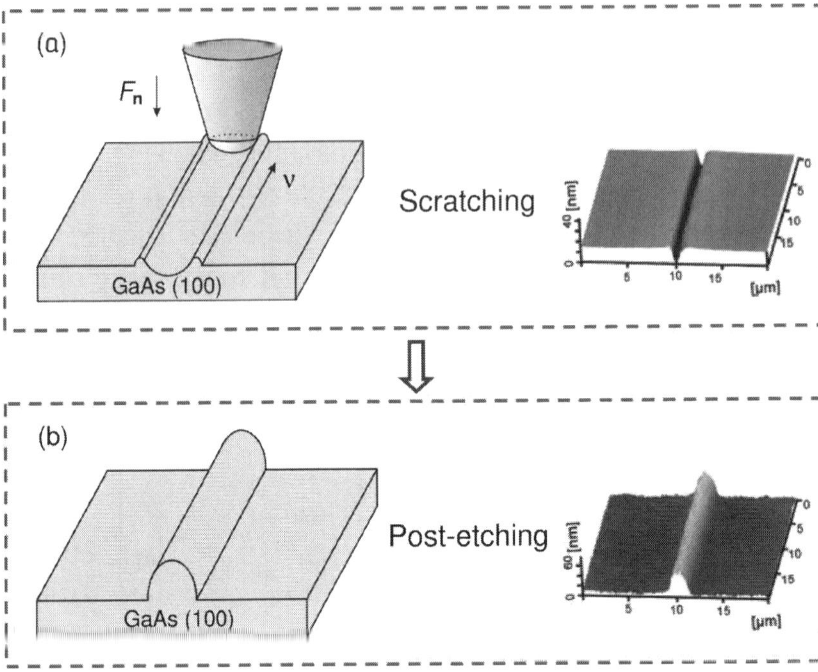

Figure 1 Schematic illustration showing the friction-induced selective etching on GaAs surface. (a) A groove was formed on GaAs surface after scratching a diamond tip under a normal load of F_n. **(b)** A protrusive nanoline was created on GaAs surface after post-etching in H_2SO_4 aqueous solution.

XPS and Raman characterization

In order to investigate the mechanism of the friction-induced selective etching process, the mesas with an area of 500 μm × 500 μm and a height of 60 nm were prepared by the homemade multi-probe instrument under a normal load of 10 mN and post-etching for 30 min. The chemical state of the fabrication area on the GaAs surface was detected by an XPS (Thermo VG250,

Thermo, Waltham, MA, USA). The microstructure of the fabrication area on the GaAs surface was measured using a Raman spectrometer (RM2000, Renishaw, Gloucestershire, UK). The excitation was supplied by the 514.5 nm Ar^+ ion laser. To avoid the random error in detection, each sample was scanned for three times.

RESULTS AND DISCUSSION

Fabrication of GaAs nanostructures

Effect of etching period on friction-induced selective etching

The etching period was found to show an obvious effect on the fabrication of GaAs nanostructures. After scratching on the GaAs surface under a normal load F_n of 20 mN, a groove with a depth of about 15 nm was created on the GaAs surface. Subsequently, a protrusive nanostructure was observed on the groove area after dipping the specimen into H_2SO_4 aqueous solution for 5 min. Figure 2 showed the AFM images and cross-sectional profile curves of the protrusive nanostructures after scratching and post-etching. The variation of the height of these protuberances with etching period was plotted in Figure 3. It was observed that the height of GaAs protrusive structure gradually increased from 12 to 94 nm with the increase in etching period from 5 to 60 min. Such results indicated that the etching rate of the scratched area was much less than that of monocrystalline GaAs. The scratched area can act as an etching mask in H_2SO_4 solution.

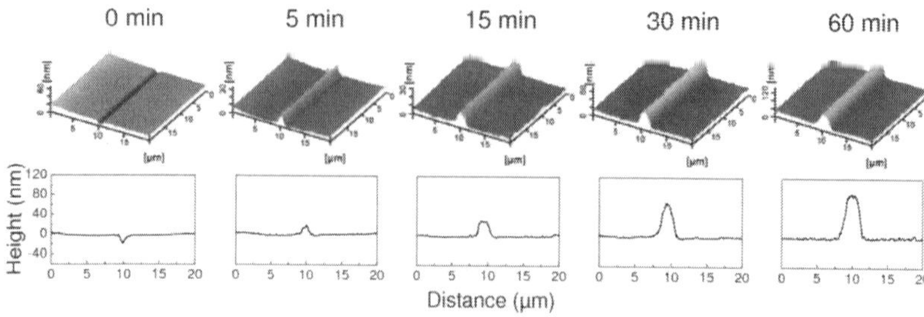

Figure 2: Effect of etching period on fabrication of GaAs surface by scratching and post-etching. The AFM images (top) and cross-sectional profiles (bottom) of the nanostructures were obtained after scratching under a normal load of 20 mN and post-etching in the H_2SO_4 aqueous solution for 5, 15, 30, and 60 min, respectively.

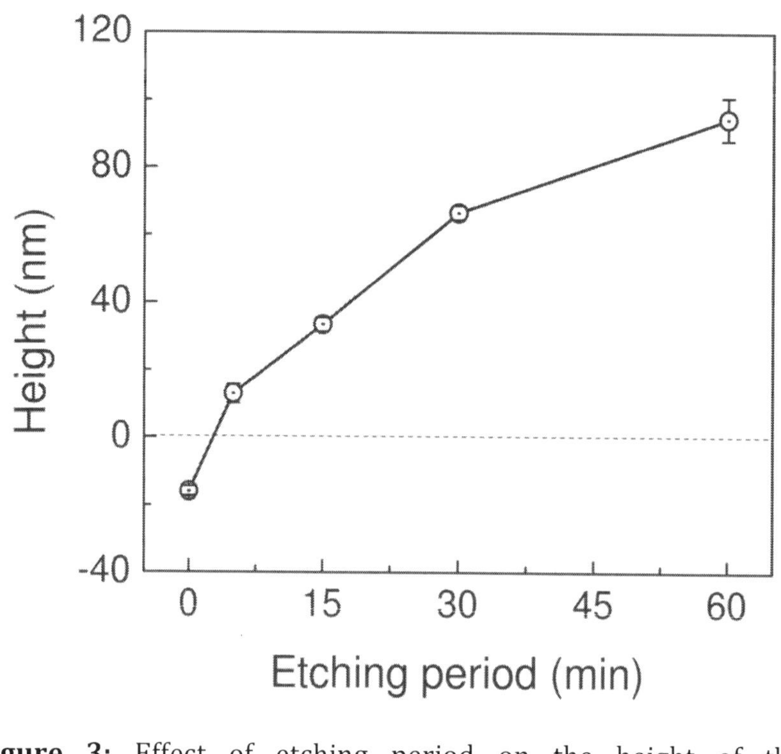

Figure 3: Effect of etching period on the height of the nanostructure on GaAs surface.

Effect of normal load on friction-induced selective etching

Aside from the etching period, the normal load also reveals an effect on the fabrication of the GaAs surface. As shown in Figure 4a, scratching tests were performed on the GaAs surface under various normal loads ranging from 0.5 to 30 mN. When the normal load was 0.5 mN, there was no obvious scratching trace on the surface. When the normal load was increased to 2 mN, a slight groove with a depth of about 0.5 nm was formed on the GaAs surface. However, when the normal load exceeded 10 mN, the scratching damage became severe and the depth of the groove increased to 23 nm at 30 mN. After etching in H_2SO_4 aqueous solution for 30 min, there was no visible etching difference on the wearless scratched surface, as shown in Figure 4b. However, the protuberance piled up gradually from the groove area when the normal load increased from 2 to 30 mN. Therefore, the critical load for the friction-induced fabrication on the GaAs surface is 2 mN, under which the Hertzian contact pressure P_c is estimated as 4.85 GPa [17, 18]. Such contact pressure was very close to the critical Hertzian contact pressure for the initial yield of GaAs surface [19]. The height of those protuberances was plotted in Figure 5. It can be seen that the height of these protuberances increased with the normal load during scratching. When the load was 30 mN, the height of nanostructures could get to 75 nm. Since the protuberance formed only in the wear area, the fabrication mechanism could be related to the deformation of the substrate induced by the mechanical interaction. The detailed generation mechanism of the protrusive nanostructures on the GaAs surface will be discussed in the next section.

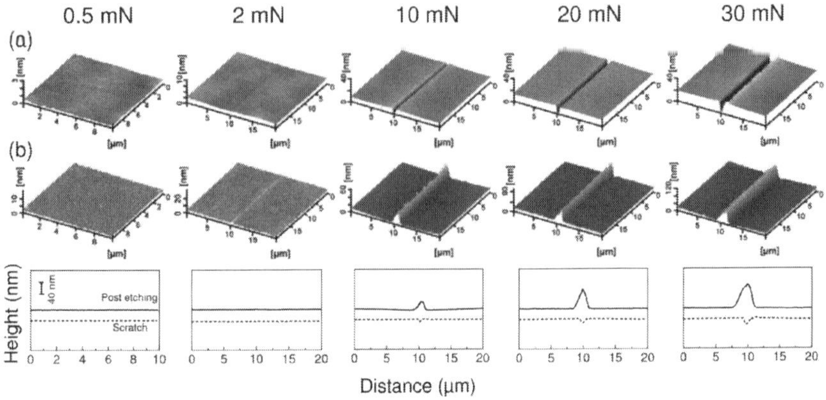

Figure 4: Effect of normal load on the fabrication of GaAs surface by scratching and post-etching. (a) AFM images of the scratches created on the GaAs surface under various normal loads. **(b)** AFM images of the nanolines on the GaAs surface after etching in H_2SO_4 aqueous solution for 30 min. The cross-sectional profiles were plotted below for the comparison.

Mechanism of the friction-induced selective etching on GaAs surface

Effect of surface oxide on the friction-induced selective etching

Extensive work has shown that various nanostructures can be produced on monocrystalline silicon and quartz surfaces by the friction-induced selective etching method [20, 21]. Guo et al. [22] suggested that both the tribochemical reaction and the transmutation of crystal structure on the scanned area can result in friction-induced selective etching. To investigate whether the tribochemical reaction played the role in the selective etching of the GaAs surface, X-ray photoelectron spectroscope was used to

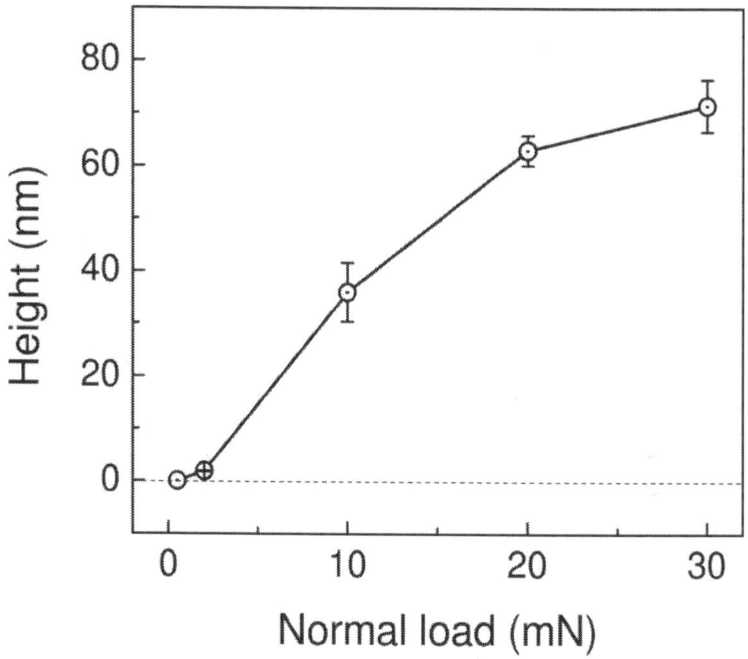

Figure 5: Effect of normal load on the height of the nanostructure on the GaAs surface.

detect the possible change of chemical composition on the original surface, scratched surface, and post-etching surface, respectively. The variation in the bonding states of Ga was presented in Figure 6. On the original surface, it was observed that there were two Ga3*d* peaks, i.e., Ga-O (Ga_2O_3) bond at 20.05 eV and Ga-As bond at 18.74 eV [23], which meant that a native oxide layer existed on the sample surface. On the scratched area, the signal of Ga-O was a little stronger than that on the original surface. After etching for 30 min, the mask effect still existed and the signal of Ga-O was very weak. Such little oxide may come from the natural oxidation of GaAs surface during the period between the finish of sample preparation and the start of XPS detection. In addition, from the X-ray full spectrum of GaAs before and after scratching, no other element or chemical compound was found in the process of the fabrication beside GaAs and its oxide. All these results confirmed that only slight tribochemical oxidation occurred on the GaAs surface during scratching. Since it was reported that the oxide of GaAs has a higher solubility into H_2SO_4 solution than GaAs substrate [24], the oxide layer may not play a role as etching mask. Therefore, the scratch-induced structural deformation was expected to act as a mask during the generation of GaAs nanostructures in H_2SO_4 solution.

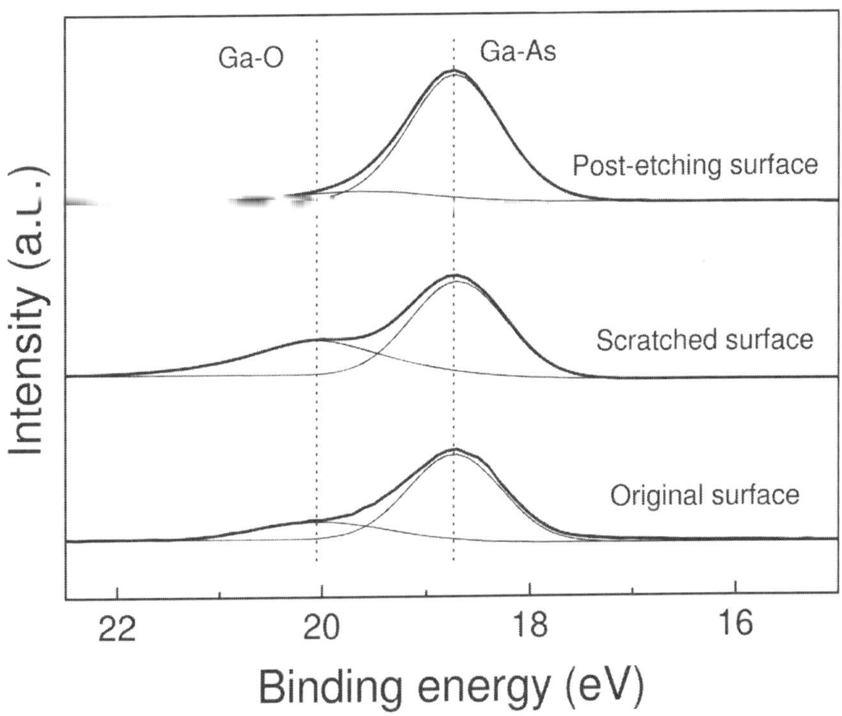

Figure 6: XPS analysis on chemical bonding states of Ga element. The detection were performed on original surface, scratched surface, and post-etching surface (scratched surface after etching) of GaAs, respectively.

Effect of structural deformation on the friction-induced selective etching

To verify whether the scratch-induced structural deformation occurred during the fabrication process, the Raman detection was conducted on original GaAs surface, scratched surface and post-etching surface. As shown in Figure 7, the Raman spectra of the original GaAs (100) displays both a longitudinal optical (LO) phonon at 290.4 cm^{-1} and transversal optical (TO) phonon at 267 cm^{-1}[25, 26]. After scratching, the LO Raman peak became wider and the positive frequency shift was 7 cm^{-1} compared to that on the original surface. When the post-etching was finished, the LO Raman peak of the mesa surface showed a negative shift of about 2 cm^{-1}. The shift and broadening of the peaks can be ascribed to the structure disorder of GaAs lattice [27]. Moreover, the positive frequency shift of LO phase is a typical character of residual compressive stress. The higher the residual compressive stress, the greater the density of crystal structure [28, 29]. As shown in Figure 8, the dense structure was expected to delay the diffusion of the etchant into internal GaAs substrate, which reduced the etching rate of the scratched area. Therefore, the dense structure can act as a 'mask' in the friction-induced selective etching of GaAs. It should be noted that compared to solely mechanical scratching, the GaAs nanostructures produced by the proposed method will have relatively lower destruction.

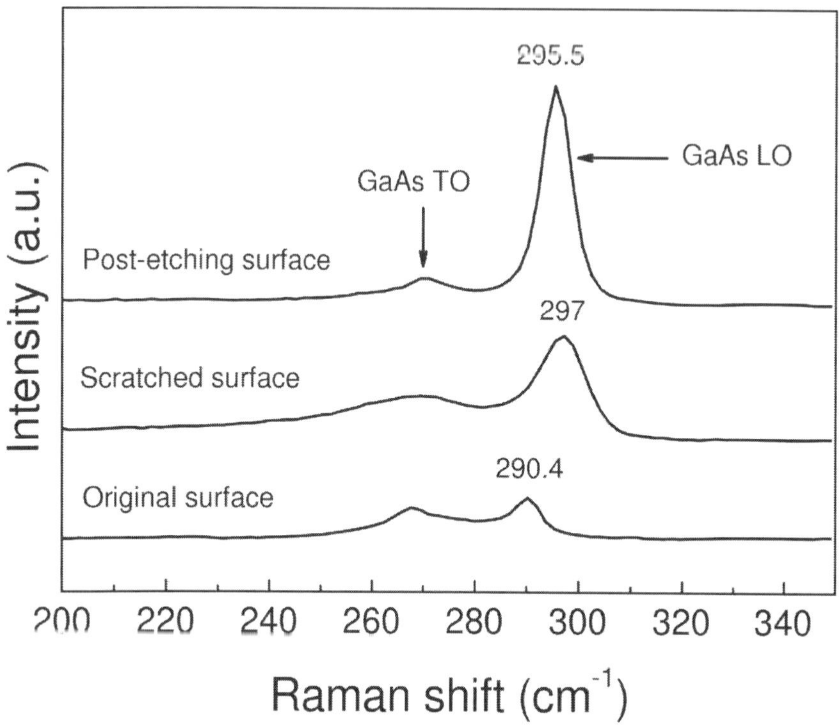

Figure 7: Raman detection on GaAs surface. The spectra were obtained from original surface, scratched surface, and post-etching surface (scratched surface after etching), respectively.

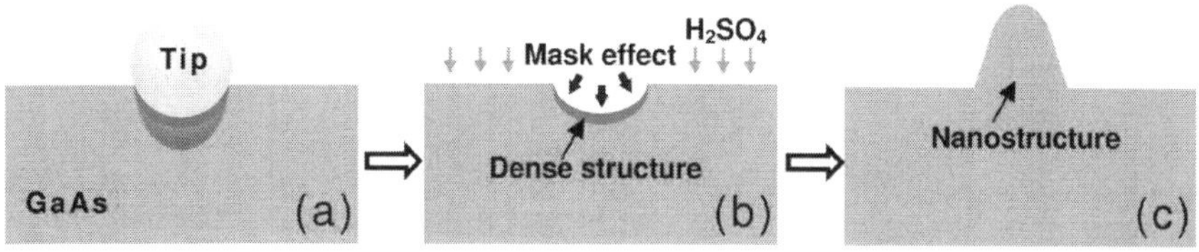

Figure 8: Schematic picture showing fabrication mechanism of GaAs nanostructure.

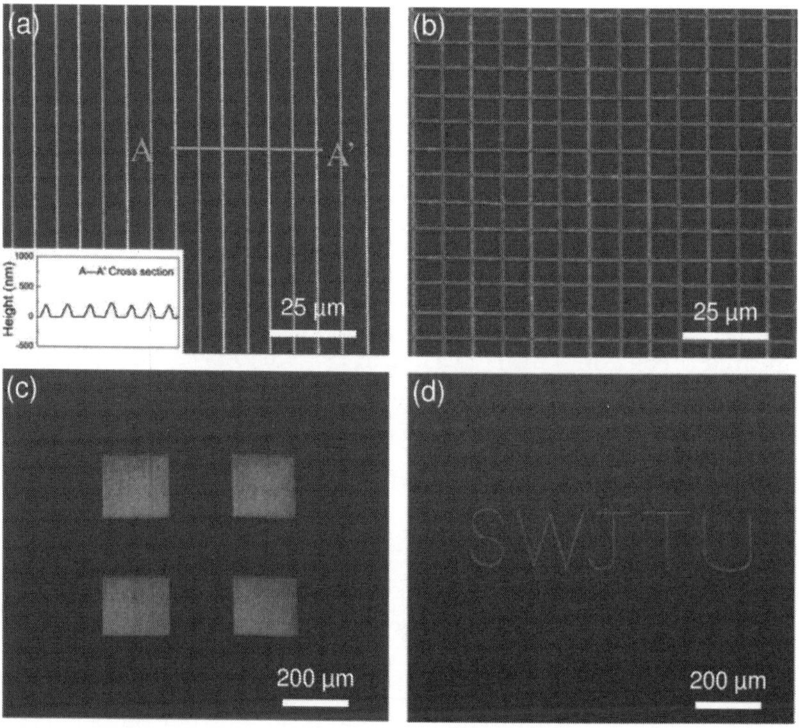

Figure 9: SEM images of GaAs patterns fabricated by friction-induced selective etching. (a) Linear arrays, **(b)** intersecting parallels, **(c)** surface mesas, **(d)** nanoletters 'SWJTU'.

Fabrication of surface pattern on GaAs surface

Based on the friction-induced selective etching method, different patterns were produced on the GaAs surface by a homemade multi-probe instrument [15]. Figure 9a displayed the linear array with the height of 200 nm and width of 2 µm. The lower left inset in Figure 9a showed the cross-sectional profile of the selected nanolines (marked by line A-A'). In Figure 9b, when the scanning traces were conducted, both on horizontal and vertical directions, intersecting parallels GaAs pattern were produced after post-etching for 2 h. The height of the GaAs nanolines was about 200 nm and the pitch width was about 9 μm. Such pattern may shed new light in orderly formation of the quantum dots or liquid drop in the manufacture process of quantum devices [30]. Figure 9c showed a 200 μm × 200 μm mesa array through continuous scanning at a normal load of 10 mN and post-etching for 1 h. In Figure 9d, the letters 'SWJTU' (short for Southwest Jiaotong University) on GaAs surface was 'written' by the scanning program control. Therefore, various patterned GaAs substrates can be achieved by controlling the normal load, scanning trace, and etching period on the GaAs surface. It is suited for large scale machining with more flexibility.

In summary, the present study proposed a friction-induced selective etching method on GaAs surface. XPS and Raman detection demonstrated that the residual compressive stress and the lattice densification was the main reason for the selective etching. Various patterns can be created on a target GaAs surface. Without any resist mask and applied voltages, this method provides a straightforward and more maneuverable micro/nanofabrication method on the GaAs surface.

CONCLUSIONS

A friction-induced selective etching method was presented to fabricate nanostructures on GaAs surface. The effects of normal load and etching period on the formation of nanostructures were investigated. The mechanism for the selective etching was discussed based on the XPS and Raman analysis. The main conclusions can be summarized as below:

1. Nanostructures can be created on the GaAs surface after scratching and post-etching in H_2SO_4solution. The height

of the nanostructures increased gradually with the increase in applied normal load or etching period.

2. Based on the XPS and Raman detection, it was found that the residual compressive stress and lattice densification induced by the scratching process were probably the main reason for the friction-induced selective etching.

Various nanostructures including line arrays and nanopatterns can be produced on the GaAs surface by the controlment of normal load, scanning trace, and etching period. Without any resist mask and applied voltages, the proposed method will open new opportunity for the micro/nanofabrication of GaAs.

ABBREVIATIONS

AFM: atomic force microscope

EB: electron beam

FIB: focused ion beam

GaAs: gallium arsenide

LO: longitudinal optical

RMS: root-mean-square

TO: transversal optical

XPS: X-ray photoelectron spectroscope

XRD: X-ray diffraction.

ACKNOWLEDGEMENTS

The authors would like to thank Prof. Zhiming Wang and Prof. Jiang Wu from the University of Electronic Science and Technology of China for their useful discussions. The authors are grateful for the support from the Natural Science Foundation of China (91323103 and 51305365) and from the Specialized Research Fund for the Doctoral Program of Higher Education of China (20130184120008).

COMPETING INTERESTS

The authors declare that they have no competing interests.

AUTHORS' CONTRIBUTIONS

PT, JG, and CS finished the fabrication experiments and acquired the original data in this article. LQ and BY have made substantial contributions to the conception and design for this article. All the authors read and approved the manuscript.

REFERENCES

1. Wu J, Shao D, Dorogan VG, Li AZ, Li S, DeCuir EA, Manasreh MO, Wang ZM, Mazur YI, Salamo GJ: Intersublevel infrared photodetector with strain-free GaAs quantum dot pairs grown by high-temperature droplet epitaxy. *Nano Lett* 2010, 10: 1512–1516.

2. Warburton RJ: Single spins in self-assembled quantum dots. *Nat Mater* 2013, 12: 483–493.

3. McNeil RPG, Kataoka M, Ford CJB, Barnes CHW, Anderson D, Jones GAC, Farrer I, Ritchie DA: On-demand single-electron transfer between distant quantum dots. *Nature* 2011, 477: 439–442.

4. Taylor C, Marega E, Stach EA, Salamo G, Hussey L, Munoz M, Malshe A: Directed self-assembly of quantum structures by nanomechanical stamping using probe tips. *Nanotechnol* 2008, 19: 015301.

5. Lee JH, Wang ZM, Liang BL, Black WT, Kunets VP, Mazur YI, Salamo GJ: Selective growth of InGaAs/GaAs quantum dot chains on pre-patterned GaAs (100). *Nanotechnol* 2006, 17: 2275–2278.

6. Gao L, Hirono Y, Li MY, Wu J, Song S, Koo SM, Kim ES, Wang ZM, Lee J, Gregory J, Salamo GJ: Observation of Ga metal droplet formation on photolithographically patterned GaAs (100) surface by droplet epitaxy. *IEEE T Nanotechnol* 2012, 11: 5.

7. Chou SY, Keimel C, Gu J: Ultrafast and direct imprint of nanostructures in silicon. *Nature* 2002, 417: 835.

8. Morita N, Kawasegi N, Ooi K: Three-dimensional fabrication on GaAs surfaces using electron-beam-induced carbon deposition followed by wet chemical etching. *Nanotechnol* 2008, 19: 155302.

9. Martin AJ, Saucer TW, Rodriguez GV, Sih V, Millunchick JM: Lateral patterning of multilayer InAs/GaAs(001) quantum dot structures by *in vacuo* focused ion beam. *Nanotechnol* 2012, 23: 135401.

10. Grenci G, Pozzato A, Carpentiero A, Sovernigo E, Tormen M: Nanofabrication of hard X-ray optics by metal electroplating in a dry etched mechanically stable inorganic template. *Microelectron Eng* 2011, 88: 2552–2555.

11. Baumgärtel T, von Borczyskowski C, Graaf H: Detection and stability of nanoscale space charges in local oxidation nanolithography. *Nanotechnology* 2012, 23: 095707.

12. Avouris P, Hertel T, Martel R: Atomic force microscope tip-induced local oxidation of silicon: kinetics, mechanism, and nanofabrication. *Appl Phys Lett* 1997, 71(2):285–287.

13. Song HZ, Usuki T, Ohshima T, Sakuma Y, Kawabe M, Okada Y, Takemoto K, Miyazawa T, Hirose S, Nakata Y, Takatsu M, Yokoyama N: Site controlled quantum dots fabricated using an atomic-force microscopy assisted technique. *Nanoscale Res Lett* 2006, 1: 106–166.

14. Hyon CK, Choi SC, Song SH, Hwang SW, Son MH, Ahn D, Park YJ, Kim EK: Application of atomic-force-microscope direct patterning to selective positioning of InAs quantum dots on GaAs. *Appl Phys Lett* 2000, 77: 16.

15. Wu ZJ, Song CF, Guo J, Yu BJ, Qian LM: A multi-probe micro-fabrication apparatus based on the friction-induced fabrication method. *Front Mech Eng* 2013, 8(4):333–339.

16. Hendrickson J, Helfrich M, Gehl M, Hu D, Schaadt D, Linden S, Wegener M, Richards B, Gibbs H, Khitrova G: InAs quantum dot site-selective growth on GaAs substrates. *Phys Status Solidi C* 2011, 8: 1242–1245.

17. Song CF, Li XY, Yu BJ, Dong HS, Qian LM, Zhou ZR: Friction-induced nanofabrication method to produce protrusive nanostructures on quartz. *Nanoscale Res Lett* 2011, 6: 310.

18. Fang TH, Chang WJ, Lin CM: Nanoindentation and nanoscratch characteristics of Si and GaAs. *Microelectron Eng* 2005, 77: 389–398.

19. Taylor CR, Malshe AP, Salamo G, Prince RN, Riester L, Cho SO: Characterization of ultra-low-load (μN) nanoindents in GaAs (100) using a cube corner tip. *Smart Mater Struct* 2005, 14: 963–970.

20. Sung IH, Yang JC, Kim DE, Shin BS: Micro/nano-tribological characteristics of self-assembled monolayer and its application in nano-structure fabrication. *Wear* 2003, 255: 808–818.

21. Song CF, Li XY, Yu BJ, Dong HS, Qian LM, Zhou ZR: Maskless and low-destructive nanofabrication on quartz by friction-induced selective etching. *Nanoscale Res Lett* 2013, 8: 140.

22. Guo J, Song CF, Li XY, Yu BJ, Dong HS, Qian LM, Zhou ZR: Fabrication mechanism of friction-induced selective etching on Si (100) surface. *Nanoscale Res Lett* 2012, 7: 152.

23. Suedu-Bob CC, Saied SO, Sullivan JL: A X-ray photoelectron spectroscopy study of the oxides of GaAs. *Appl Surf Sci* 2006, 183: 126–136.

24. Ghidaoui D, Lyon SB, Thompson GE, Walton J: Oxide formation during etching of gallium arsenide. *Corrosion Sci* 2002, 44: 501–509.

25. Zardo I, Yazji S, Marini C, Uccelli E, Morral AF, Abstreiter G, Postorino P: Pressure tuning of the optical properties of GaAs nanowires. *ACS Nano* 2012, 6(4):3284–3291.

26. Gotoshia SV, Gotoshia LV: Laser Raman spectroscopy of phase transformation in GaAs induced by radiation defects. *Phys Status Solidi C* 2013, 4: 646–649.

27. Pizani PS, Lanciotti F, Jasinevicius RG, Duduch JG, Porto AJV: Raman characterization of structural disorder and residual strains in micromachined GaAs. *J Appl Phys* 2000, 87: 1280.

28. Attolini G, Francesio L, Franzosi P, Pelosi C, Gennari S, Lottici PP: Raman scattering study of residual strain in GaAs/InP heterostructures. *J Appl Phys* 1994, 75: 4156.

29. Champagnon B, Martinet C, Boudeulle M, Vouagner D, Coussa C, Deschamps T, Grosvalet L: High pressure elastic and plastic deformations of silica: In situ diamond anvil cell Raman experiments. *J Non-Cryst Solids* 2008, 254: 569–573.

30. Kiravittaya S, Heidemeyer H, Schmidt OG: Growth of three-dimensional quantum dot crystals on patterned GaAs (001) substrates. *Phys E* 2004, 23: 253–259.

2

CHAPTER

Selective Plasma Etching of Polymeric Substrates for Advanced Applications

Harinarayanan Puliyalil [1,2,†] and Uroš Cvelbar [1,2,*,†]

[1]*Jožef Stefan International Postgraduate School, Jamova cesta 39, 1000 Ljubljana, Slovenia*

[2]*Jožef Stefan Institute, Jamova cesta 39, 1000 Ljubljana, Slovenia*

ABSTRACT

In today's nanoworld, there is a strong need to manipulate and process materials on an atom-by-atom scale with new tools such as reactive plasma, which in some states enables high selectivity of interaction between plasma species and materials. These interactions first involve preferential interactions with precise bonds in materials and later cause etching. This typically occurs based on material stability, which leads to preferential etching of one material over other. This process is especially interesting for polymeric substrates with increasing complexity and a "zoo" of bonds, which are used in numerous applications. In this comprehensive summary, we encompass the complete selective etching of polymers and polymer matrix micro-/nanocomposites with plasma and unravel the mechanisms behind the scenes, which ultimately leads to the enhancement of surface properties and device performance.

Keywords: plasma etching; selectivity; surface chemistry; nanomesh; nanostructuring; polymer composite; biomaterials; adhesion; etch mask; radical atoms

1. INTRODUCTION

Plasma technology is one of the fastest developing branches of science, which is replacing numerous conventional wet-chemical methods in high-tech laboratories and industries, with a huge impact in renewable energy, environmental protection, biomedical applications, nanotechnology, microelectronics, and other fields. Plasma, the complex mixture of ions, radicals, electrons, and excited molecules has replaced conventional methods to develop various nanostructured materials with complex morphology and advanced properties (e.g., the production of vertically aligned carbon nanotubes (CNTs), which is difficult to achieve with other synthetic methods [1,2]). Owing to fast multi-scale production, it is a preferred method for the synthesis of other nanomaterials such as nanowires (NWs), carbon nanowalls (CNW), graphene sheets, *etc.* [3,4,5]. Through this, plasma in nanoscience also impacts the field of renewable energy and environmental protection, where we endeavor to replace hydrocarbon energy resources with solar energy or hydrogen fuel cells [6,7]. An additional benefit of plasma technology is the wet free doping of semi-conducting nanomaterials with heteroatoms to alter the band gap energy and conductivity for various applications [8,9,10]. In the overview of the applications in biology and medicine, plasma is found in various applications, from plasma surgery and the manufacture of artificial implants to the straightforward disinfection of the medical equipment.

The main processing routes in plasma technology are simply concluded into deposition and etching. For achieving the best material performance, these two techniques are either used individually or in a combination. Plasma-enhanced chemical vapor deposition (PECVD) is a technique complementary to thermally excited chemical vapor deposition (CVD) and is used for creating thin layer coatings, micro- or nanostructures, and even the deposition of complex functional materials [11]. Various applications of plasma deposition techniques are found in numerous articles, and some examples are given in [12,13,14,15]. On the other hand, plasma etching originates from the interactions between plasma particles with various substrates. The interactions are either physical or chemical; the particles with high kinetic energy are utilized to knock out the atomic or molecular species from the surface, whereas in the latter case the interaction of particles is purely potential. The physical interactions are mostly linked to ionized plasma species, while potential is linked to neutral or excited species.

Plasma etching, referred many times as plasma chemical etching or dry etching, of both organic and inorganic materials was reported for material fabrication in multidisciplinary applications. Compared to the wet chemical etching, plasma etching is capable of controlled and precise etching at very small scales (~10 nm). Additionally, plasma etching limits the disadvantages, for example, via contamination or solvent absorption during the treatment process. Some of the major advantages and disadvantageous of wet chemical etching *versus* plasma etching are listed in Table 1. In this review, we will mostly focus on the plasma etching of organic materials, which typically undergo etching relatively faster than inorganic materials. The difference in the etching arises from the stability of materials towards various chemical species. As an example, when we compare with liquid chemistry, alkali metals react explosively with water or acids, whereas transition or inner transition metals react very slowly or even stay inert. In the same way, different organic or inorganic materials react at different rates with various plasma species. This advantage of the disparate reactivity of substrates towards plasma is then utilized for fabricating important micro-/nanostructured materials.

Table 1. Comparison of wet chemical etching *versus* plasma etching.

	Wet Chemical Etching	Plasma Etching
Etchant:	Chemical (acids, alkali, *etc.*).	Reactive gas (radicals, ions, *etc.*).
Etch rate and selectivity:	High.	Good, controllable.
Advantageous:	Low equipment cost, fast processing and easy to implement.	Capable of small scale etching (~10 nm), no contamination issues, no hazardous chemicals, ecologically benign technology.
Disadvantageous:	Inadequate to define small feature size less than 1 μm, handling of hazardous chemicals, contamination issues, ecologically unfriendly technology with need of waste processing.	High equipment cost, implementation dependent on application, potential radiation damage.
Directionality:	Only isotropic etching.	Can be isotropic or anisotropic.

In the field of material fabrication, the organic materials, especially polymer materials, are replacing many inorganic substrates. Simple fabrication and low production costs are the two great advantages for employing polymer materials. However, to minimize certain disadvantages of pure polymers, their properties are improved by making composites from multiple organic materials using multiplication of material properties or even reducing their dimensions to the nanoscale.

To achieve this, the preferential removal of materials is desired, which is simply achieved by plasma etching. This approach was found to be the best method for many applications to achieve desired surface structures, morphology, or chemistry [16,17]. One should stress the effects on the surface chemistry, especially when it comes to organic materials, as it is typically more important than the morphology for certain applications such as adhesion. Plasma etching always alters the surface energy by the successful incorporation or formation of chemical groups during the interaction of plasma particles with materials [18].

Up to now, no reviews have reported specifically on the topic of plasma selective etching, which has innumerable applications in the field of materials science. This review concentrates on the applications of selective etching. However, prior to that, a short discussion is made on the important types of plasmas based on particle species, their energy, and the result of interaction with the surface, namely sputtering, reactive ion etching, or neutral radical chemical etching. All these interactions then drive the representative applications of plasma etching schematically presented in Figure 1. Sputtering, which involves the removal of various substrates merely by high-energy particle collision is not very important for the selective etching of materials. On the contrary, the reactive ion etching and the neutral plasma chemical etching are found to be efficient for the preferential removal of one material over the other. Following this, the applications of the plasma modified polymeric substrates are described. The main objective of this article is to focus on the characteristics of polymer materials, which are decisive for the etching rate and stability towards high-energy particles. Plasma processing of various polymer composites and their advanced applications are outlined as well.

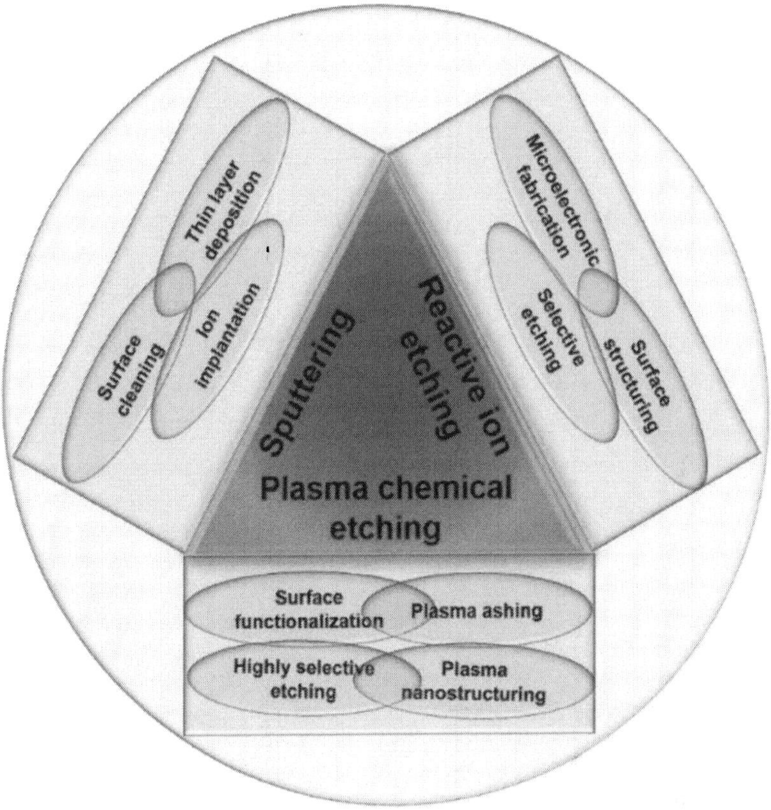

Figure 1. Schematic representation of various uses of different plasma processes.

2. PLASMA PROCESSING

2.1. Sputtering

Sputtering deposition is widely employed as a robust method in a thin film preparation for countless applications. The basic principle of the sputtering technique is the generation of high-energy ions, which are then accelerated and guided towards the substrate where they knock out surface atoms as ions into the vapor phase. The so-generated ions are also guided and deposited on the coated substrate where they form a thin layer. However, this deposition process is beyond the scope of this paper. More interesting are energetic surface collisions, where material is removed from the surface.

While considering the process, one should create an ion-rich environment that enables the acceleration of generated particles by applying an external electric or magnetic field. The number density and kinetic energy of the ions (N_{ion} and $W_{K.E.}$, respectively) should be sufficiently large to overcome the energy barrier created by the binding energy of the surface atoms ($W_{B.E.}$). The sputtering process is efficient when the kinetic interactions significantly prevail over chemical and thermal interactions, which typically occurs at ion energies ranging higher than several 100 eV or even higher than 1 keV [19]. However, the sputtering process efficiency in atom removal is significantly dependent upon the material sputtered. One simple schematic representation of the plasma-sputtering chamber is presented in Figure 2a.

The major drawback of the sputtering process is its inefficiency for the surface modification of delicate materials including polymers. The problem is that the high-energy particle–surface interactions are uncontrollable and completely damage the surface mostly through cascade collisions inside the material, which provide almost no selectivity. Moving away from high-energy ions, which cause only the pure kinetic-physical interactions of ions with surfaces (via two body collisions), there is a gray zone of so-called chemical sputtering, where some chemical reactions between the incident beam and substrate occur. This process involves mostly lower energy ions (typically with energies around 100 eV) and is commonly applied to remove the surface contaminates such as hydrocarbons [20].

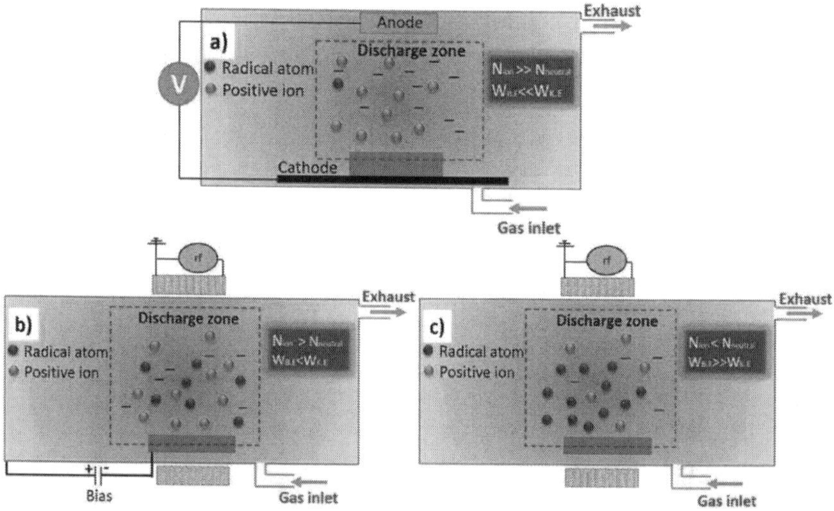

Figure 2. Schematic representation of various plasma processing systems for (**a**) sputtering; (**b**) reactive ion etching; and (**c**) highly dissociated weakly ionized plasma for chemical etching.

2.2. Reactive Ion Etching

Reactive ion etching (RIE) was mostly developed and improved by the microelectronic industry in the last few decades for trenching and patterning the surfaces. The RIE process is very diverse and uses various plasma combinations, with species ranging from merely low-energy ions (10 eV to several 100 eV) to combinations of ions with radicals, excited atoms, and electrons. These combinations are mostly generated with glow discharges, where the ratio of ($W_{K.E.}$) to ($W_{B.E.}$) is much lower than in the case of sputtering (Figure 2b). In these processes, the etching material is typically DC-biased in order to increase the effect of ion collection and acceleration towards the surface. The process is most commonly done inside a chamber with low pressure of a selected gas, where plasma is generated in a high-frequency discharge [21]. In such systems, where the surface potential is controlled, the interacting ions are directed, their energy distribution controlled by bias voltage. Nevertheless, other reactive species including neutrals inside the plasma play a significant role in determining the etching rate and etching anisotropy. Like in the case of any plasma process, the etching rate inside a reactive ion chamber is partially dependent on the discharge parameters that influence plasma properties. Therefore, the increase in the discharge power improves the etching rate, since the dissociation efficiency increases with the applied power. On the other hand, the generation of self-bias potential at reduced pressure improves the etching efficiency by energizing the particles [22].

The most reported discharge systems used for RIE are capacitively coupled RF plasma (CCP) and inductively coupled RF plasma (ICP). In CCP plasma, the ion energy varies as a function of applied power, whereas an increase in gas pressure reduces the acceleration of ions by collisions with the background gas. The ion energy flux to the surface is increased by bias voltage, which imparts a high kinetic energy to the particles. This makes complications by reducing the control or selectivity over the etching process. However, lowering the pressure also results in an increase in the collision mean free path, which restricts the existence of plasma at the point in which the mean free path approaches a value that is of the same order than the distance between the electrodes [22,23]. On the other hand, ICP is capable of generating high particle densities without any additional biasing, even at higher gas pressures [24].

The etching rate in RIE processing varies from a few nm to a few hundred nm per minute depending on the conditions used for the processing [25]. Optimization of the process is extremely important because of the problems encountered during RIE such as bowling, undercutting, or mask scattering, all of which adversely affect the aspect ratio of the processing, especially during the fabrication of circuit boards [26]. To improve the etching rate and aspect ratio, RIE is sometimes combined with other processing techniques including magnetic-field-induced beam control, lithography, ultraviolet (UV) irradiation, *etc.* [27,28].

2.3. Neutral Radical Plasma Chemical Etching

Very frequently, processes like ion sputtering and even RIE lead to thermal damages of the surface or even of bulk properties, which make them improper for the etching of temperature-sensitive materials such as polymers. In order to avoid overheating by ion bombardment or even by increased neutral gas temperature, it is wise to use mostly neutral plasma species like neutral atoms. This is achieved with so-called cold plasmas mostly generated at lower pressures (around 100 Pa) with high-frequency discharges. The high frequency applied accelerates the electrons, whereas the heavier ions cannot follow frequency, which results in a high dissociation and low ionization of the feeding gas (Figure 2c). When treatments are performed in the post-glow or after-glow regions, the discharge features practically no ions. Microwave discharges are popular for the plasma treatment of materials because they are characterized by low plasma potential, which has the advantage of insignificant ionic effects, but the neutral gas temperature is normally high. On the other hand, RF discharges have the electron temperature of about 5 eV for plasma gases such as O_2, where the observed neutral gas temperature is lower compared to microwave discharge. In such high frequency discharges, the dissociation is almost linearly enlarged with the increase in discharge power, independent of the gas flow [29].

Highly dissociated plasma is not only applicable for polymer etching, but also for other applications such as chemical reduction, nanostructuring, and plasma cleaning [30,31,32]. The reactions of polymers and other materials will be more elucidated in the following chapters, where different neutral atoms contribute a major part to etching through adsorption and recombination processes [33,34,35]. The recombination of atoms at preferential spots on the surface originates from the roughness of the surface and plays a crucial role, even in deciding the nanostructure growth [24,36]. The surface interactions and bond-breaking mechanisms during the interactions of non-equilibrium plasmas with various carbonaceous materials are also reported [37,38].

3. APPLICATIONS OF PLASMA FUNCTIONALIZATION AND ETCHING OF POLYMERS

Polymer materials cover a large segment of our material requirements for packaging, microelectronics, photonic devices, medical implants, sealing applications, water repellent coatings, thermal and electrical insulators, sensing materials, *etc.* [39,40,41,42]. Plasma-assisted surface functionalization and etching were proven to provide the desired surface energy and morphological changes to the polymeric surface [43,44]. Although various chemical treatments such as soaking in acidic medium are efficient and inexpensive methods for improving the surface energy of polymeric materials, they are associated with persisting residuals after treatment and, to a large extent, connected to environmental pollution [45,46]. Plasma modification consists of surface functionalization, which is considered to be a primary step in which surface chemistry is altered by bond breaking and the incorporation of functional groups. Thereafter, the removal of material takes place through the bonding of surface atoms with impinging radicals, which recombine and leave the surface. Through this mechanism, the organic surface contaminants or weakly bonded surface layer are removed. An alternative to plasma pre-treatment is UV irradiation, which mostly results in surface functionalization, but it is not sufficient for material etching. Moreover, wet chemical treatments as the second alternative are limited by the type of incorporated functional groups [47]. By taking these factors into consideration, plasma treatments are preferably used in polymer treatments for a wide variety of applications.

Among various applications of plasma surface treatments, improving the adhesion of various metals on polymer substrates is an important task. The major challenge in metallizing polymeric surfaces is their lower surface energy, which confronts the adhesion of metallic particles. By incorporating suitable polar functional groups by plasma exposure, this deficiency is easily solved. For instance, Pascu *et al.*presented both microwave and RF plasma treatments of polyvinylidene fluoride with N_2 and NH_3 as working gases for the incorporation of polar nitrogen containing groups including amines, nitriles, or imines. These incorporated polar functionalities then aided to increase the amount and adhesive strength of the deposited copper [48]. Metallized polymers including polyethylene terephthalate (PET) are important in the fabrication of microelectronic and photonic devices, where the intermixing of the metal and polymer showed the significant improvement of their properties when material was exposed to post-deposition annealing in Ar or Ar/O_2 plasma [49]. The treatment of the samples was done with Ar/O_2 plasma generated at low pressures (50–100 Pa) within RF discharge at power 35 W. The process increased the surface roughness and porosity, allowing the diffusion of coated Au/Ag metal from the surface into the bulk. To obtain the best adhesion, the plasma treatment conditions were optimized to minimize fracture at the interface between the metal film and polymer substrate [50,51]. The adhesion of Cu layer on polyamide also increased with O_2 plasma treatment at 200 W and 8 Pa for 3 min prior to Cu metallization. The peel strength of the deposited Cu on polyamide increased to a value of 250 g/mm, which was almost 280 times higher than that of the non-treated sample [50]. Irrespective of the gas or discharge condition used for the treatment, the metal coatings of polymers should be done immediately after the plasma surface modification in order to eradicate the drawbacks arising from the aging effects [52,53].

In addition to the improved adhesion of inorganic particles, plasma treatments have a significant role in fabricating surfaces, which are customized for the adhesion of various organic or biomolecules. Surface treatment with oxidizing plasma source gases such as O_2, NH_3, N_2, Air, or even noble gas mixtures is found efficient to enhance the surface energy and thereby adhesion of various bio materials onto the surface [54,55]. The surface roughness and polarity provided by plasma exposure are utilized to increase the adsorption of anti-blood clotting agents such as heparin onto various polymeric substrates. This operation is extended in application for manufacturing artificial organs [56]. In addition to this, the cell growth on various polymeric surfaces efficiently increases after suitable plasma pre-treatments as a simultaneous effect of improved surface contact area and the electron rich behavior. The incorporated functional groups allow for the nucleation and adsorption of desired biomaterials on the surface of the implant used by which the spatial alignment of the bio-molecules on the surface is regulated [57,58]. Another supportive application of plasma etching in the bio-medical field is plasma sterilization of surgical devices, drug packaging, or processing of other medical objects [59,60]. For example, on exposure to reactive O_2 plasma neutral radical in the afterglow, the blood proteins and microorganisms showed exponential etching rate, whereas the PET substrate exhibited only a linear etching rate [61]. These distinct etching rates help to remove toxins or other organic contaminants from the surface without affecting the surgical equipment or medical material. Further applications of plasma degradation of materials extend into novel applications such as decontamination of toxic warfare agents, the cleaning of dental devices, and even the precise removal of cancerous cells [62,63,64].

Besides the improved surface energy of polymeric substrates, plasma treatment is also able to incorporate low-energy functionalities on the polymeric surface. This is largely applied in the fabrication of water-repellent surfaces, which are significantly important in terms of generating better anti-aging or anti-corrosive properties of membranes for oil-water separation, and of fluid transportation control [65,66]. The self-cleaning ability of materials is expressed in terms of the water contact angle, where the surface with a contact angle above 150° is termed as the superhydrophobic surface [67]. Plasma etching efficiently creates superhydrophobic surfaces by two major mechanisms: firstly by providing sufficient roughness to the surface by the etching process and secondly by providing a sufficient number of low-energy functional groups [68]. Typically used fluorine containing plasmas have an ability to simultaneously incorporate sufficient amounts of low-energy functional groups as well as create the desired surface roughness through etching. In the cases of fluorinated polymers including Teflon (polytetrafluoroethylene—PTFE), even fluorine-free gases are sufficient enough to generate hydrophobicity merely by increasing the surface roughness by polymer etching [66,69]. An example of this are nanocone structures on Teflon surface achieved after etching the surface with O_2 plasma (25 Pa and 50 W) for 10 min. In this case, the water contact angle achieved was 134°. Such nano-featured surface is achieved also by etching anisotropy, which is created with deposits like polymer beads. The surface decorated with polystyrene (PS) beads with a 10-μm diameter prior to the plasma process can create nanocones during etching as well (Figure 3) [69]. The sufficient roughness is boosted by the additional deposition of Au nanoparticles on top of the nanocones, which leads to a superhydrophobic surface response. In certain cases, the roughness on the etched surface is influenced by the sputtered deposits from the plasma chamber wall, which act as a mask to protect the polymer but provide similar anisotropy for etching [70].

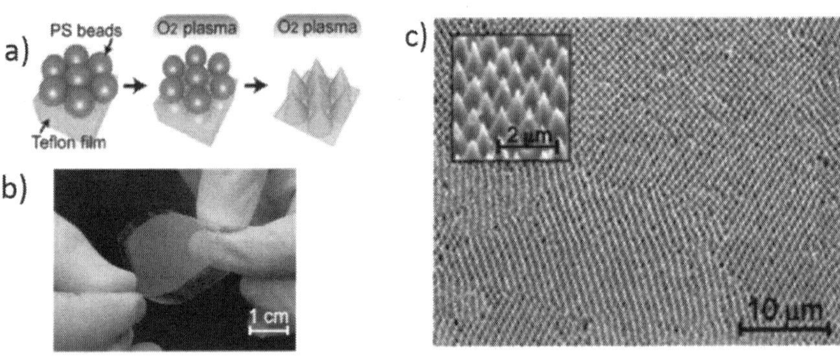

Figure 3. (**a**) Scheme of the fabrication process of Teflon nanocone arrays; (**b**) Photograph showing a macroscopic view of flexible Teflon nanocone array; (**c**) Scanning electron microscopy (SEM) images of the tilted nanocone array. Inset: detailed view of Teflon nanocones (Reproduced with permission from [69]. Copyright American Chemical Society, 2011).

4. THE ORIGIN OF PLASMA ETCHING SELECTIVITY

The importance of the dry etching of polymeric materials for numerous applications such as masking or nanostructuring has increased the demand for systematic studies on the stability of various polymeric materials towards reactive plasma particles. The stability of the polymer definitely depends on the type of polymer/monomer units present (homo or copolymer, aliphatic or aromatic, crystalline or amorphous, *etc.*) as well as the energy and type of plasma particle interacting with it. Herein, the important factors are directly related to the stability of polymers towards plasma particles, which result in selective etching that will be presented and discussed.

One of the primary structural features that are considered is the aliphatic and aromatic moiety on the polymer backbone. The etching rates for aromatic polymers are relatively lower compared to aliphatic ones due to the extra energy stabilization (~36 kcal/mole), provided by the aromaticity of the rings in the polymeric chain. One more explanation to support the distinct etching rate is that the aliphatic rings degrade easily to form volatile molecules, whereas the aromatics generally form more non-volatile fragments and are not easily removed from the surface [71]. However, plasma-initiated degradation can occur on the bond–ring junctions, which are connected through relatively unstable secondary or tertiary carbon atoms. This argument can question the extra stability of the aromatic polymer chains exposed to the plasma. An appropriate argument was given by Taylor and Wolf, who stated that the extra stability of the aromatic polymers, irrespective of the position of the aromatic ring, originated from the ability of aromatic rings to quench the reactive plasma particles [72]. The quenching is typically directed through the abstraction of the hydrogen from the aromatic ring by the incoming reactive atom and the formation of a functionalized ring (Figure 4). This reaction prevents the formation of any active sites or dangling bonds on the polymer chain and prevents the chain cleavage. Additionally, the energy released during this process will be readily reduced by contributing to the vibrational energy of the polymer backbone [73]. Some early reports on the gas phase reactions of aromatic rings with oxygen plasma radicals significantly support this argument [74,75]. During such reactions, the hydroxylation was mostly directed towards the ortho position regardless of the size of the substituents on the ring due to the higher degree of neucleophilicity of oxygen radicals. This observation points out that the reactions are mostly guided through a kinetic route to give rise to reaction products that are thermodynamically less stable [74].

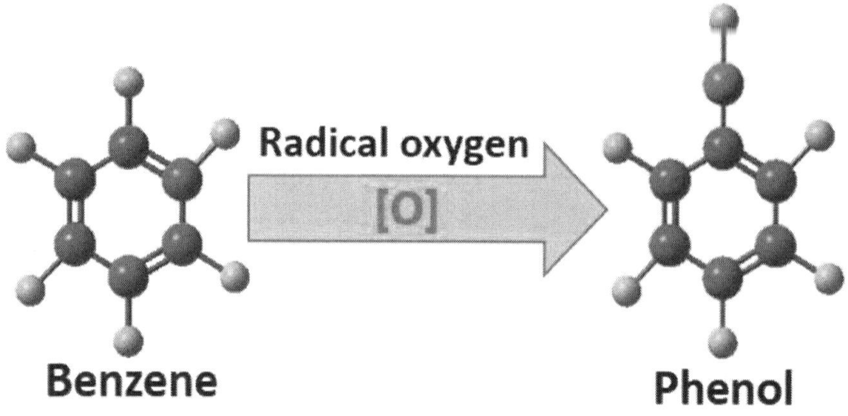

Figure 4. The schematic of reaction involved in the radical quenching by the aromatic ring to form functional group instead of ring cleavage.

Liming *et al.* provided an excellent example to present the etching efficiency and selectivity of plasma towards various chemical bonds [76]. The stability of various bonds was accurately distinguished by high-energy hydrogen radicals and ions, used to selectively etch away the edge atoms of graphene sheets and to convert them into graphene nanoribbons (GNR). As observed from atomic force microscopy (AFM) imaging, the observed etching rate for single layer graphene was 0.27 nm/min, whereas multi-layer graphene disclosed an etching rate of 0.1 nm/min (Figure 5). By keeping the surface temperature to an optimum value of 300 °C, the defect-free hydrogen terminated graphene nanoribbons were synthesized. The selectivity originated from the lower chemical stability of the edge carbon towards plasma particles compared to the atoms embedded in the mid-region of the graphene sheet [77]. In the same manner, while considering the polymeric substrates, regardless of the etching mechanism, the etching rate was mostly connected with the energy required for breaking the polymer backbone, as demonstrated by Taylor and Wolf. They compared the etching rate of different polymers (k_{rel}) with the number of chain scissions per 100 eV of energy absorbed (G_s) (Table 2). The etching rate displayed a linear relationship with the bond-breaking energy of the polymer backbone [72].

Another important parameter which determines the etching selectivity is the energy required for the initiation of chemical reaction (activation energy). For example, the chemical oxidation of Mg in air is highly exothermic. However, the reaction will not commence unless sufficient heat is provided such that the reactants can reach the intermediate transition state (activated complex). Once the reaction starts, the exothermic energy released will be used for the reaction to proceed. The comparison of activation energies required for the reaction between n-butane with various neutral species in the vapor state discloses that O (^{1}D) and F atoms do not need any activation energy for the chemical reaction to start, which reduces the selectivity towards various hydrocarbon bonds.

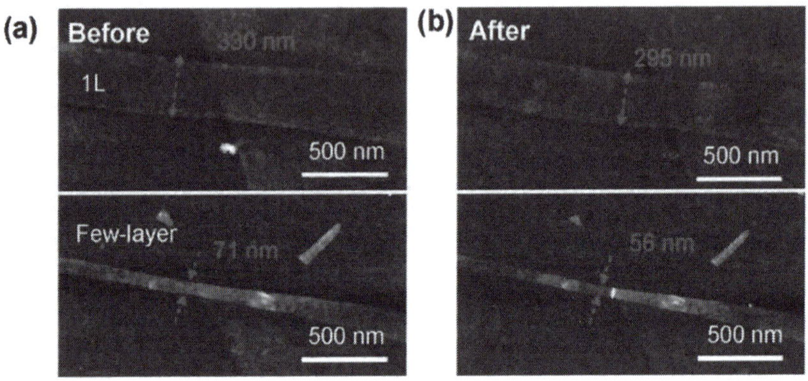

Figure 5. Atomic force microscopy (AFM) images of two small pieces of graphene (top: a monolayer (1 L) graphene strip; bottom: a few-layer graphene strip). (**a**) Before and (**b**) after selective hydrogen plasma edge etching for 60 min at 300 °C (Reproduced with permission from [76]. Copyright American Chemical Society, 2010).

Table 2. Relative rates of O_2 plasma removal (k_{rel}) and G_s-values for selected polymers (Reproduced with permission from [72]. Copyright John Wiley and Sons, 2004).

No.	Polymer	k_{rel}	G_s
1	Poly(α-methylstyrene)	1.11	0.3
2	Polyphenyl methacrylate	1.33	—
3	Polyviny1 methyl ketone	1.48	—
4	Polymethy1 methacrylate (PMMA)	2.37	1.2
5	Polymethyl methacrylate-co-methacrylonitrile (94:6 mol %)	2.70	2.03
6	Polyisobutylene	3.56	4
7	Polybutene-1 sulfone	7.11	8

Although O (^{3}P) species are reactive, the energy of activation is found to be 1 eV to react with the hydrocarbon, which indicates that this species can distinguish various chemical bonds much better than O (^{1}D) or F species [73,78]. Likewise, the Cl atomic species are highly reactive without any bond selectivity towards the reacting polymer due to no activation energy barrier. On the other hand, Br atoms are relatively non-reactive due to an energy barrier of around 1.8 eV. Moreover, the etching rate can also be controlled through the combination of various reactive species, which can be regulated by controlling the feeding gas composition and discharge parameters. In one recent report, it was demonstrated that adding O_2 into CF_4 plasma increased the etching rate [79]. This phenomenon was due to the involvement of excited oxygen radicals in the electron impact dissociation of CF_4, which leads to a higher dissociation and more F atomic species [80].

Hegemann *et al.* compared the etching rates for various polymers with different chemical structures inside the noble gas-generated plasmas and concluded that the polymer backbones that contained oxygen functionalities etched away relatively faster. For example, polymethyl methacrylate (PMMA), PET, *etc.* showed relatively higher etch rates compared to the hydrocarbon polymers inside Ar plasma generated at 300 W and a pressure of 20 Pa (Figure 6) [81]. The explanation for the higher etch rates for functionalized polymers was due to the secondary reactive oxygen atomic species released into the system as a result of polymer bond cleavage. More surprisingly, polypropylene (PP) exhibited a lower etching rate compared to PS, which was in contradiction to the hypothesis of radical quenching mechanism proposed for the extra stability of aromatic polymers. However, this was explained by the efficient cross-linking on the PP surface upon exposure to plasma due to the side chain activation [81].

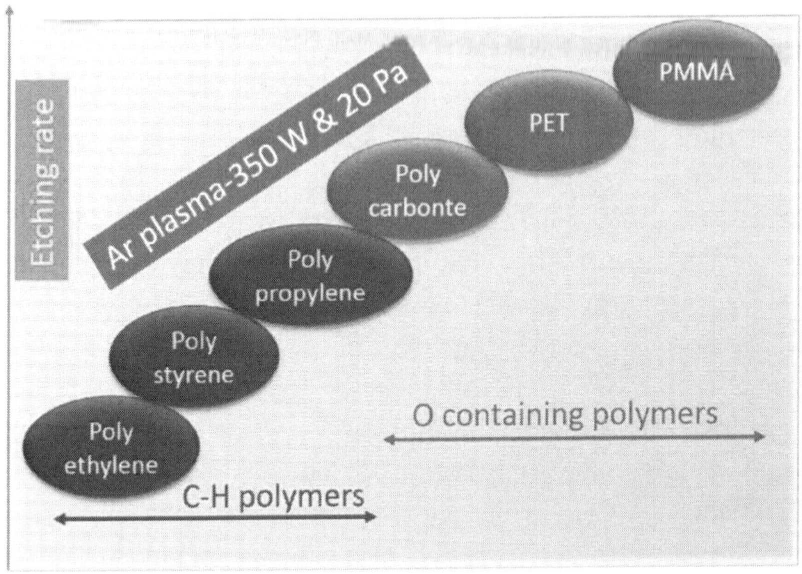

Figure 6. Etching rates for various polymer substrates in Ar plasma based on [81].

Plasma surface interactions have a strong correlation with the crystallinity of the polymer used, where the etching rate is typically reduced with the increase in crystallinity [82]. The reason for this can be different neutral atom recombination probabilities on surfaces. For example, the neutral O atom recombination probability is 8.3×10^{-4}–2.1×10^{-4} for amorphous PET or 4×10^{-4}–2.5×10^{-5} for semi-crystalline PET [83]. Thus, the extension of interactions between the polymer and the neutral atomic species that induce etching differ with respect to crystallinity, which cannot hold for charged species. The ion flux to the surface is independent of the crystallinity. The increased value of the neutral atom recombination coefficient provided higher contribution to the surface temperature which induced more deformation to the amorphous material, as reflected by the roughness measurements before and after plasma exposure [83,84,85]. The treatment of LDPE (low density polyethylene) and HDPE (high density polyethylene) inside CF_4 plasma revealed similar results. The etching rate for LDPE was 1.5 nm/min more than that of the HDPE inside CF_4 plasma generated by capacitively coupled RF discharge, where the morphological changes also displayed significant differences with respect to crystallinity [86]. Nair*et al.* recently reported the removal of an amorphous carbon layer from the multiwalled carbon nanotube (MWCNT) with low-energy ions, radicals and metastable species, where the crystalline phase remained relatively intact. The extra stability provided by the crystallization energy additionally supported the etch-resistant behavior of crystalline polymers [87]. From all of the above-mentioned results, the relative etching rates for various polymeric substrates inside the widely used oxygen plasma are summarized in Figure 7.

An important class of polymers that needs to be separately addressed is that of block copolymers, since they are largely used in block copolymer lithography for the semiconductor industry. The chemical dissimilarities between the two polymer domains permit the selective etching of one component over the other, when a suitable treatment method is used. After the selective etching of the less stable polymer, the unaffected component will form the template pattern. One of the most widely used copolymers is the PS/PMMA system, where both of the components have distinct chemical stabilities. For achieving spherical and cylindrical structures, PMMA is easily removed with a suitable wet chemical treatment [88]. However, the wet

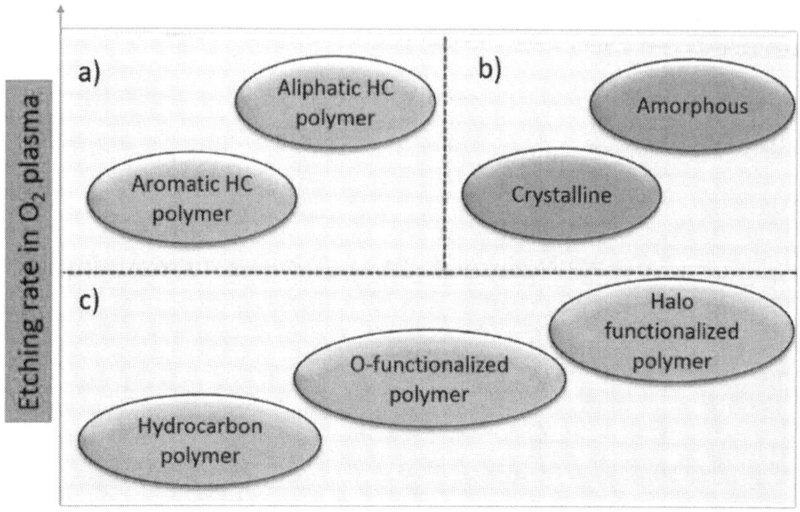

Figure 7. General schematic of the etching rate for various types of polymeric materials in O_2 plasma. (**a**) Dependence of etching rate on the aliphatic/aromatic behavior of the monomer units; (**b**) Etching rate dependence on the crystallinity; (**c**) Functionality dependence of the polymer with etching rate.

chemical etching is limited due to capillary forces when the polymers possess lamellar arrangement. In order to avoid these problems, plasma selective etching is implemented and used for a number of copolymer systems: for example, polystyrene–poly ferrocenylisopropylmethylsilane diblock copolymer, styrene–butadiene–styrene (SBS) triblock copolymer, polystyrene–polydimethylsiloxane block copolymer, polyhydroxybutyrate-co-hydroxyvalerate, *etc.* [89,90,91]. The most fundamental object while choosing the copolymer for plasma etch patterning is the selection of proper blocks that have distinct etching rates towards various plasma particles [92].

The oxidation probability of the polymer is independent of the density of plasma particle species in the proximity of the samples, but the etching still depends on plasma particle properties and their fluxes to surface [93]. In many cases, the etch-resistant properties of polymer materials are dependent on the type of discharge used, where the difference in the etching rate originates from the energies and flux of various reactive species including neutral atoms and ions generated [37,94,95]. The etching rate is also well correlated to the surface temperature originated from the ion bombardment, neutral atom recombination, and exothermic carbon oxidation [96,97,98]. In order to achieve the highest etching rates, a higher surface temperature is always preferred. However, the surface temperatures above the glass transition temperature of the polymer can critically affect the bulk properties. Due to this, the radical and ion flux should be optimized, whereas in many cases the pulse mode plasma treatment over continuous mode will be more appropriate to control the temperature related effects [99,100,101].

5. APPLICATIONS OF PLASMA SELECTIVE ETCHING OF COMPOSITE MATERIALS

Plasma selective etching is utilized in many applications for material fabrication especially in micro- and nano-structuring. One of the important examples is the transplantation of nanostructures like nanotubes. Li *et al.* demonstrated a method for growing CNTs inside Si trench with Ni catalysts. The challenge faced was the transfer of the designed nanotubes onto suitable receptors. This difficulty was solved by extracting the nanotubes with an epoxy matrix in the form of pellets, followed by oxygen plasma selective etching of the epoxy matrix in order to release well aligned and densely packed CNT arrays [102]. By taking advantage of the difference in etching selectivity of various chemical bonds towards plasma, the metal single-walled carbon nanotube (m-SWCNT) were also obtained. One of the challenges for creating m-SWCNTs is the higher chemical etching rate of m-SWCNTs compared to semi-conducting single-walled carbon nanotubes (s-SWCNT) [103]. However, by taking advantage of the weaker stability of smaller diameter s-SWCNT due to the C–C bond bending, a mild hydrogen plasma treatment yielded m-SWCNT from a mixture of m- and s-SWCNT [104]. This etching selectivity is only diameter-dependent, unlike the m- or s- character of SWCNT, and could yield a 100% recovery and easy scaling up of one type of SWCNT from the mixture [105].

Graphene based electrodes are used in wide variety of semiconducting devices such as field effect transistors, sensors, supercapacitors, *etc.* [106]. In the transistor applications, one of the most important parameters is the gap dimension between the electrodes. Graphene sheets, which are separated by micro- or nanoscale distance, played a significant role in the fabrication of numerous electronic devices. The fabrication of such devices is extremely hard by wet chemical treatment due to the lack of precision in etching over nanoscale dimensions. As demonstrated by Liao *et al.*, plasma-assisted mask etching can provide precise control over the gap between graphene sheets ranging from a few micrometers to hundreds of nanometers [107]. Furthermore, the electrical properties of graphene sheets are highly dependent on their specific surface area. For this, graphene sheets can be converted into different porous forms including graphene nanomesh, crumpled graphene, folded graphene, and graphene foam in order to exploit the exceptionally large surface area [108,109,110]. Such high surface area graphene sheets are also easy to obtain by means of polymeric or metallic mask-assisted plasma selective etching. In a presented example, the spin coating of PMMA on graphene-supported silica, when covered with porous anodized aluminum oxide resulted in a suitably masked surface for plasma etching. Treatment of this composite with O_2 plasma for 30 s resulted in pores with diameters of 67 nm on the graphene surface, which produces the graphene mesh with a very high surface area of 100 μm^2 [111]. The origin of etching selectivity during the processing was based on the tolerance of the Al mesh towards plasma compared to the PMMA layer. The observed difference in the etching rates of PS and P(S-r-MMA-r-GMA) (a random copolymers of styrene, methyl methacrylate, and glycidyl methacrylate) was 1.17 and 1.42 nm/s, respectively, inside O_2 plasma (50 W, 10 sccm, 1.3 Pa), compared to 0.76 and 0.96 nm/s, respectively, inside the CHF_3/O_2 gas mixture (300 W, 45 sccm CHF_3 and O_2 5 sccm, 8 Pa). Exploiting this, the patterned copolymer template was fabricated, which provided an easy route for creating nanoperforations on graphene sheets. Such prepared graphene with a large surface area has been used in electronic and sensing devices [112]. Additionally, the combination of holography and photoresist-assisted O_2 plasma selective etching was demonstrated for providing a low-cost production route to synthesize graphene nanomesh [113]. Additional examples are also seen where self-assembled masks are used for the perforation of graphene sheets during selective etching [114]. Such masks provide controlled exposure of the graphene layer to the plasma reactive species and yield precise mesh sizes and ribbon widths between successive meshes. This allows control over the electronic characteristics of graphene [115]. During the etching of such graphene composite surfaces, the selective edge oxidation and O_2 physisorption on the inner walls of the mesh add secondary benefits to chemo-sensing applications [116].

Another particular application of selective plasma modification and etching was to improve gas permeation properties of composite membranes [117]. After plasma treatment with corona discharge, the polyamide 6/polyethersulfone (PA6/PES) composite membrane exhibited significant improvement in separation of CO_2, N_2, and O_2 by incorporating functional groups and increasing the surface roughness [118]. Unfortunately, the changes induced to the distinct composite components were not studied individually. When a polybutadiene/polycarbonate (PB/PC) membrane was exposed to the Cl-containing plasma, it was observed that the modified surface chemistry had very little influence, whereas the separation towards N_2 and O_2 were dominantly affected by the variation in surface morphology [119]. Other than for permeation applications, selective plasma-assisted mask O_2/Ar etching of Nafion membrane has been used in the preparation of patterned metal-polymer composite membranes for robotics and other applications [120].

Unzipping CNTs under plasma exposure is known to ease the preparation of novel materials including hollow CNTs or GNR. To make such preparative methods easier, the simplest method is to expose CNTs embedded in a polymer matrix to reactive plasmas. Exposure of the vertically aligned CNT in the polymer matrix to plasma opened the tips of the embedded CNTs. This route also included the removal of the polymer matrix at higher rates and then exposing the tube tips and further opening of the CNT tips [121]. In another relevant example, plasma selective etching enabled the successful synthesis of single-layer GNR from CNTs embedded in PMMA composite. The method provided excellent control over edge smoothness and uniformity in width for the fabricated ribbon structures [122].

Due to its simplicity, and the higher rate of dimension tolerance compared to wet chemical etching procedures, plasma etching is well exploited in the semiconducting industry for printing integrated circuit boards. One of the disadvantages often given for plasma processing is the high production cost due to expensive low-pressure plasma systems, which can be bridged only with long-term operation. Silicon wafers, as one of the most used materials in the semiconductor industry, is efficiently etched mostly by halogen-containing plasmas or by their mixture with other gases in the presence of a suitable etch mask. The etch mask on the surface generates an unbalanced etching rate of the surface and creates desired patterns on the wafer [123]. For achieving the etching selectivity between the semiconducting layer and the photoresist, the control of the plasma species inside is needed. By adding certain gases, it is possible to scavenge the radicals and ions, which presents an alternative route to controlling the species inside the plasma and the corresponding etching selectivity, without changing the discharge parameters [124]. One of the major drawbacks of RIE technology is the non-uniform etching on the sidewalls. It operates in such a way that the radicals get accumulated nearer to the edge of the wafer instead of the center, which increases the etch rate towards the edges. Additionally, the ion current is favorably drawn over the wafer edge, and the etch rates for the silicon wafer show an increase from the center to the edge and the consequent edge breakings [125]. Performance level and many of the disadvantages of RIE processing such as anti-notch performances, quality of the profile, surface finishing, *etc.* are tuned by adjusting the applied frequency or pulsing [125]. Even in the presence of these disadvantages, plasma etching is far better than other conventional methodologies.

An important application of the diverse material etching rates in plasma is used to deal with the nanostructuring of the polymeric substrates. Polymer-based 1D and 2D nanostructures have potential applications in sensing and energy devices [126,127]. Although template-assisted synthetic methods are well-known, controlling the dimensions during template removal still represents an obstacle. Plasma-assisted mask etching has proven to be an alternative to overcome these problems. An example is the exposure of metallic nanoparticles deposited on polymer surfaces to reactive gas plasma, which yields the dense polymeric NWs. The metallic coating or, say, metallic material islands (nanodots) merely acted as an etch mask to provide a rough surface to mobilize the NW growth [128]. For this purpose, a mixture of Ar, O_2, and CF_4 gases was leaked into the inductively coupled plasma chamber at constant flow ratios of 15, 10, and 30 sccm. The plasma discharge was generated at 400 W where an additional 100 W was used for biasing to accelerate and direct the ions to the surface. In this way, the polymeric NWs with an aspect ratio of up to 700 were produced on the surface of a large number of polymers including PET, Kapton, Dura film, PS, and polydimethyl siloxane (PDMS). The length of the NWs exhibits a linear trend with the exposure time, whereas the diameter remains constant at a value around 100 nm for a given thickness of the mask on the surface. On the other hand, the density of the NWs is strongly influenced by the thickness of the coated metal. The scanning electron microscopy (SEM) images of NW with respect to various thicknesses of the mask and the relationship of the etching rate with the plasma exposure time are presented in Figure 8.

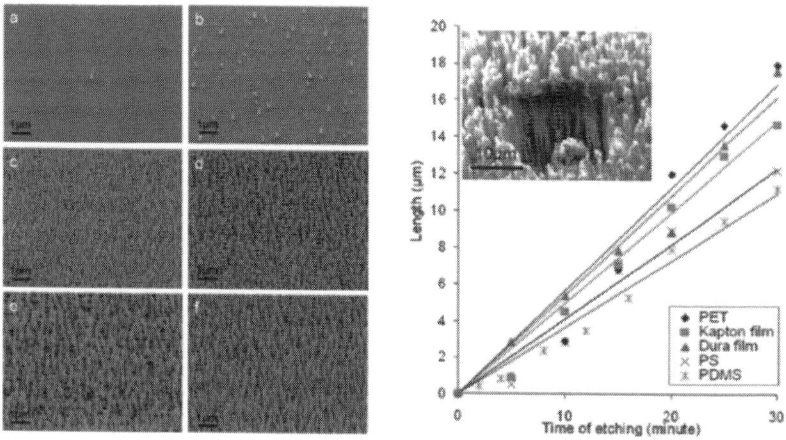

Figure 8. SEM images of the density-controlled fabrication of polymer nanowire (NW) arrays of Kapton by covering the initial surface with (**a**) 0.75; (**b**) 1.5; (**c**) 3; (**d**) 4.5; (**e**) 10; and (**f**) 15 nm of Au before inductively coupled plasma (ICP) etching. The graph represents the length-controlled growth of NWs of polyethylene terephthalate (PET), Kapton film, Durafilm, polystyrene (PS), and polydimethyl siloxane (PDMS). The inset is a SEM image of a NW array on Durafilm after 30 min of etching (Reproduced with permission from [128]. Copyright American Chemical Society, 2009).

In many of the presented examples, the surface nanostructuring is achieved by using an appropriate mask/template. Nevertheless, it was not necessary to introduce a mask prior to processing. A typical example is found where simultaneous plasma-enhanced reactive ion synthesis and etching (SPERISE) is employed. This technique was first used during the treatment of a Si wafer inside HBr-O_2 gas plasma (Figure 9a). In the primary step, the halogen species interact with the surface, and Si atoms are released into the vapor phase. This excited Si is then combined with the O and Br reactive species to form an etch-resistant silicon oxy bromide complex, which is deposited back on the surface. These deposits then act as masks and support the non-uniform etching on the surface. As a result of this processing, the mushroom-like Si NWs are formed, where an etch mask is left on the top (Figure 9b). The removal of the etch mask is done later with relatively simple mild chemical treatment [129].

Another application of plasma selective etching includes the study of filler dispersion inside the composite matrix. The filler dispersion inside the polymer matrix is the deciding factor for its bulk properties including ductility, hardness, impact resistance, *etc.* Conventional methods such as cryogenic breaking are generally used to monitor the dispersion of the fillers [130,131]. However, this method lacks precision for the determination of dispersion due to the possible displacement of the fillers during the distortion of the matrix. To overcome this, dissociated plasma etching is applied where the surface polymer layer is selectively removed to expose the embedded fillers. A fast and selective removal of the surface polymer from the composite materials including powder coatings and paint films with the dissociated oxygen plasma followed by SEM imaging enables the monitoring of the filler dispersion [132,133,134,135]. The same strategy allowed for the optimization of the bonding process efficiency of powder coatings with other materials, where bonding is enabled through disclosed but still embedded fillers. Due to the inhomogeneity in the plasma etching of various components

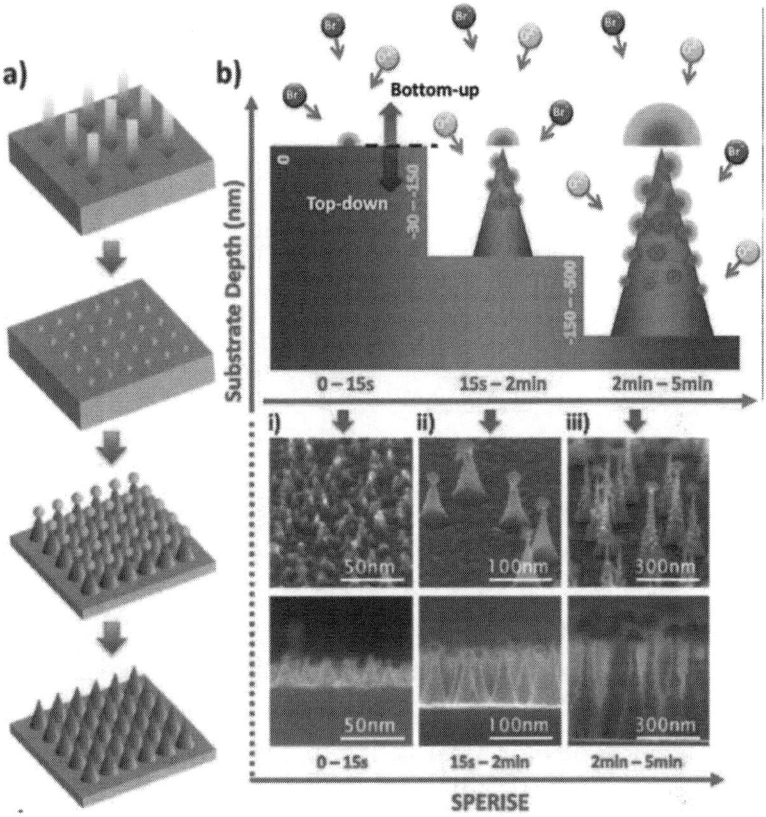

Figure 9. Simultaneous plasma enhanced reactive ion synthesis and etching (SPERISE) process and Si nanocone formation mechanism. (**a**) Process flow of the nanomanufacturing process: Pseudo randomly distributed silicon oxybromide nanodots are synthesized on the planar silicon substrate surface in the first few seconds of the SPERISE process. The oxide nanodots grow to hemispheres by a phase-transition nucleation process and act as a protective nanomask for the simultaneous reactive ion etching of the silicon underneath. Depending on the growth rate of the oxide hemispheres and the crystalline structures of the silicon substrates, nanocones with different aspect ratios are formed. The silicon oxybromide nanohemispheres on top of the

nanocones are removed by wet etching; (**b**) Detailed schematic drawing of the three typical stages in the SPERISE process: Bromine and oxygen reactive ions interact with silicon to form synthesized oxide hemisphere and dots (orange) and etched silicon cone structure (green). Both the illustrations and corresponding SEM images at (i) 0–15 s; (ii) 15 s–2 min; and (iii) 2–5 min in the SPERISE process manifest this unique nanomanufacturing method (Reproduced with permission from [129]. Copyright American Chemical Society, 2011).

inside the composite material, the resulting surface acquires relatively high surface roughness due to the exposed fillers. Along with suitable plasma functionalities, increased surface roughness is gained, which additionally improves the metallization of the surface [136,137]. Similarly, the fabrication of counter electrodes for solar cell application was done from a PP/CNT nanocomposite, where O_2 plasma reactive ion etching was used to remove the thick protruding layer of PP on the surface to expose the embedded CNTs. As a result, the charge transfer resistance of the composite surface is diminished to much lower values [138].

The applications of plasma selective etching are further extended for achieving improved bio- and chemo-sensing properties of semiconducting materials. As shown in one recent report, the gas-sensing properties of a composite membrane prepared by electrospinning process from the aq. solution of polyvilnyl alcohol (PVA) and SnO_2 was improved by O_2 plasma treatment. The resulting SnO_2 material showed a notable sensing response towards very low concentrations of ethanol vapor (~1 ppb). The enhanced sensitivity was featured because of the high specific surface area of the ripple-like structures obtained after plasma treatment followed by annealing at 500 °C [139]. However, the metal oxide nanodevices are less preferred due to their inability to operate at room temperature for sensing applications that could be replaced by conducting polymers, carbon allotropes or their suitable composites [140,141,142]. The sensing ability of such materials, especially in the context polymeric composites are further improved by plasma modification of the surface. As demonstrated by Raghu and co-workers, plasma modifications of the MWCNT composite with conducting polymers indicated greater effects on sensitivity and selectivity towards various volatile organic compounds [143]. However, a better understanding of how plasma improves the sensitivity and selectivity is still to be studied extensively for improving the sensor properties. While in the case of bio-sensors, the sensitivity is found to be controlled by the covalent attachment of the bio-molecule to the sensing material. For this application, the removal of the polymer matrix and the functionalization of the exposed MWCNTs on the surface of a pristine PS/MWCNT composite yielded an opportunity to covalently attach antibodies and to fabricate advanced immunosensors [144].

One of the most recent applications of plasma chemical etching was the improvement of the insulation properties of polymeric composites. The insulating properties of various polymeric materials are standardized in terms of comparative tracking index (CTI) as per international electrotechnical commission (IEC) grading. The origin of poor insulating properties of composite materials is an aftereffect of the polymer charring on the surface at a high voltage electric arc. This issue is commonly tackled by adding suitable fillers, which forms little or no char on the surface [145]. However, this is inadequate due to the thin polymer layer on top of the embedded fillers. The simplest solution to overcome this situation is the selective removal of the surface polymer layer by an optimized plasma discharge. The cold O_2 plasma removal of the surface polymer from the glass embedded phenolic resin composite was demonstrated to improve CTI performances up to 56% (Figure 10) [101]. The performance level increased with the decrease in the surface polymer content along the prolonged plasma treatment times.

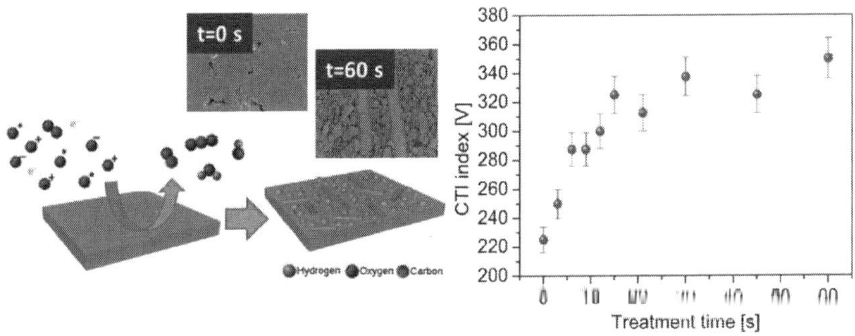

Figure 10. Plasma surface interaction of the glass-filled composite with corresponding SEM images for non-treated and plasma-treated samples for 60 s. The graph represents the variation of comparative tracking index (CTI) performance with plasma exposure time (Reproduced with permission from [101]. Copyright Royal Society of Chemistry, 2015, Year.").

6. CONCLUSIONS AND RESEARCH CHALLENGES TO TACKLE

This review attempted to outline the interdisciplinary applications of plasma etching and the selective etching of polymer-based materials from different branches of science. Moreover, it looks at the origin of selectivity and attempts to find answers bringing together an understanding of plasma properties with very diverse results of plasma–surface interactions through the note to nanomaterials. Among various plasma modifications, neutral dense and ion free cold plasmas are have been found to be efficient, especially for treating delicate polymeric and biomaterials to avoid unwanted surface damages and thermal effects. The plasma-induced functionalization and etching of polymer substrates is preferred to wet chemical etching and UV irradiation for designing the surface chemistry, surface morphology, and surface energy. These plasma induced modifications enable the attachment of various materials and biomolecules onto the surface. Additionally, plasma-induced hydrophobization of the surface can effectively increase the water-resistant behavior of the surface to improve anti-aging, anti-fouling, and corrosion-resistant properties. Such improvements in the surface properties are directly connected to structural and morphological changes.

The high-energy particles inside the plasma are able to distinguish various bond types under controlled process parameters. Utilizing the difference in the chemical stability of various bond types on the same molecule, delicate materials such as GNRs have been fabricated. The comparison of etching rate for different classes of polymeric materials has revealed that the difference in the etch rate is connected with both physical and chemical properties of the material. The presence of aromatic moieties has reduced the etching rate by radical quenching, whereas the surface cross-linking has reduced the etching rate for branched polymers. Generally, the etch rates are lower for hydrocarbon polymers compared to functionalized macromolecules. Additionally, a higher extension of crystallinity has reduced the plasma etch rates for various polymers, presented through reports on a few examples, namely, PET, LDPE, and HDPE. Thus, existing differences in the etching rates for various polymeric substrates have been employed in the surface structuring of block copolymers in applications, especially for the semiconductor industry. The pronounced applications of plasma selective etching of composite materials have been found to be efficient for reducing the dimensionalities of materials such as CNT or graphene sheets for advanced applications in electronics, electrical, and sensing devises. Furthermore, the preferential etching of composite surfaces has been effectively used for a simple and low-cost synthesis of polymer NWs with a controlled aspect ratio.

The plasma etching rate and etching selectivity of micro/macro molecules are well connected with the strength of the available bonds in the material. However, the side reactions such as radical quenching and surface cross-linking significantly affect the etching rate. Thus, the optimization of the process parameters is essential for achieving etching selectivity in different systems. Plasma is used as a single operating tool in only one/a few step(s) throughout multistep processes for demanding applications. More frequently used plasma has been found as one of the steps in multi-processing. This is clearly visible in many of the discussed examples, including the synthesis of GNR, graphene nanomesh, patterning the surface of block copolymers, *etc.* In some of the presented examples, the involved wet chemistry had slight adverse effects on the properties of the final product. An additional challenge in front of the plasma community is the pinpoint control of various plasma species and their reactions on the atomic or molecular level. Such developments are important, especially for such remarkable applications as plasma nanoscience and plasma medicine.

ACKNOWLEDGMENTS

This work was partially funded by the Slovenian Research Agency (ARRS) projects L2-6769 and program P2-0082. Harinarayanan Puliyalil would like to thank G. Filipič for valuable discussions, and the Jožef Stefan International Postgraduate School (MPŠ) for the grant from the Innovative Scheme for Research.

CONFLICTS OF INTEREST

The authors declare no conflict of interest. The founding sponsors had no role in the design of the study; in the collection, analyses, or interpretation of data; in the writing of the manuscript, and in the decision to publish the results.

REFERENCES

1. Ostrikov, K.; Cvelbar, U.; Murphy, A.B. Plasma nanoscience: Setting directions, tackling grand challenges. *J. Phys. D* **2011**, *44*.

2. Zhang, Z.; Han, S.; Wang, C.; Li, J.; Xu, G. Single-walled carbon nanohorns for energy applications. *Nanomaterials* **2015**, *5*, 1732–1755.

3. Bo, Z.; Yang, Y.; Chen, J.; Yu, K.; Yan, J.; Cen, K. Plasma-enhanced chemical vapor deposition synthesis of vertically oriented graphene nanosheets. *Nanoscale* **2013**, *5*, 5180–5204.

4. Filipič, G.; Cvelbar, U. Copper oxide nanowires: A review of growth. *Nanotechnology* **2012**, *23*.

5. Guo, S.; Dong, S. Graphene nanosheet: Synthesis, molecular engineering, thin film, hybrids, and energy and analytical applications. *Chem. Soc. Rev.* **2011**, *40*, 2644–2672.

6. Ashik, U.P.M.; Wan Daud, W.M.A.; Abbas, H.F. Production of greenhouse gas free hydrogen by thermocatalytic decomposition of methane—A review. *Renew. Sustain. Energy Rev.* **2015**, *44*, 221–256.

7. Mariotti, D.; Mitra, S.; Svrcek, V. Surface-engineered silicon nanocrystals. *Nanoscale* **2013**, *5*, 1385–1398.

8. Park, S.H.; Chae, J.; Cho, M.-H.; Kim, J.H.; Yoo, K.-H.; Cho, S.W.; Kim, T.G.; Kim, J.W. High concentration of nitrogen doped into graphene using N_2 plasma with an aluminum oxide buffer layer. *J. Mater. Chem. C* **2014**, *2*, 933–939.

9. Meena, J.S.; Chu, M.-C.; Chang, Y.-C.; You, H.-C.; Singh, R.; Liu, P.-T.; Shieh, H.-P.D.; Chang, F.-C.; Ko, F.-H. Effect of oxygen plasma on the surface states of ZnO films used to produce thin-film transistors on soft plastic sheets. *J. Mater. Chem. C* **2013**, *1*, 6613–6622.

10. Pumera, M. Heteroatom modified graphenes: Electronic and electrochemical applications. *J. Mater. Chem. C* **2014**, *2*, 6454–6461.

11. Kumar, A.; Ann Lin, P.; Xue, A.; Hao, B.; Khin Yap, Y.; Sankaran, R.M. Formation of nanodiamonds at near-ambient conditions via microplasma dissociation of ethanol vapour. *Nat. Commun.* **2013**, *4*.

12. Attri, P.; Arora, B.; Choi, E.H. Utility of plasma: A new road from physics to chemistry. *RSC Adv.* **2013**, *3*, 12540–12567.

13. Meyyappan, M. A review of plasma enhanced chemical vapour deposition of carbon nanotubes. *J. Phys. D* **2009**, *42*.

14. Meyyappan, M. Plasma nanotechnology: Past, present and future. *J. Phys. D* **2011**, *44*.

15. Vasilev, K.; Griesser, S.S.; Griesser, H.J. Antibacterial surfaces and coatings produced by plasma techniques. *Plasma Process. Polym.* **2011**, *8*, 1010–1023.

16. Aria, A.I.; Lyon, B.J.; Gharib, M. Morphology engineering of hollow carbon nanotube pillars by oxygen plasma treatment. *Carbon* **2015**, *81*, 376–387.

17. Coburn, J.W.; Winters, H.F. Ion- and electron-assisted gas-surface chemistry—An important effect in plasma etching. *J. Appl. Phys.* **1979**, *50*, 3189–3196.

18. Denes, F.S.; Manolache, S. Macromolecular plasma-chemistry: An emerging field of polymer science. *Prog. Polym. Sci.* **2004**, *29*, 815–885.

19. Bohlmark, J.; Lattemann, M.; Gudmundsson, J.T.; Ehiasarian, A.P.; Aranda Gonzalvo, Y.; Brenning, N.; Helmersson, U. The ion energy distributions and ion flux composition from a high power impulse magnetron sputtering discharge. *Thin Solid Films* **2006**, *515*, 1522–1526.

20. Smith, T. Sputter cleaning and etching of crystal surfaces (Ti, W, Si) monitored by auger spectroscopy, ellipsometry and work function change. *Surf. Sci.* **1971**, *27*, 45–59.

21. Oehrlein, G.S. Reactive-ion etching. *Phys. Today* **1986**, *39*, 26–33.

22. Lee, H.; Oberman, D.B.; Harris, J.S. Reactive ion etching of GaN using CHF_3/Ar and C_2ClF_5/Ar plasmas. *Appl. Phys. Lett.* **1995**, *67*, 1754–1756.

23. Levchenko, I.; Keidar, M.; Xu, S.; Kersten, H.; Ostrikov, K. Low-temperature plasmas in carbon nanostructure synthesis. *J. Vac. Sci. Technol. B* **2013**, *31*.

24. Ostrikov, K.; Levchenko, I.; Cvelbar, U.; Sunkara, M.; Mozetic, M. From nucleation to nanowires: A single-step process in reactive plasmas. *Nanoscale* **2010**, *2*, 2012–2027.

25. Tachi, S.; Tsujimoto, K.; Okudaira, S. Low-temperature reactive ion etching and microwave plasma etching of silicon. *Appl. Phys. Lett.* **1988**, *52*, 616–618.

26. Coburn, J.W.; Winters, H.F. Conductance considerations in the reactive ion etching of high aspect ratio features. *Appl. Phys. Lett.* **1989**, *55*, 2730–2732.

27. Cybart, S.A.; Roediger, P.; Ulin-Avila, E.; Wu, S.M.; Wong, T.J.; Dynes, R.C. Nanometer scale high-aspect-ratio trench etching at controllable angles using ballistic reactive ion etching. *J. Vac. Sci. Technol. B* **2013**, *31*.

28. Zeze, D.A.; Cox, D.C.; Weiss, B.L.; Silva, S.R.P. Lithography-free high aspect ratio submicron quartz columns by reactive ion etching. *Appl. Phys. Lett.* **2004**, *84*, 1362–1364.

29. Mozetič, M.; Vesel, A.; Monna, V.; Ricard, A. H density in a hydrogen plasma post-glow reactor.*Vacuum* **2003**, *71*, 201–205.

30. Filipič, G.; Baranov, O.; Mozetič, M.; Ostrikov, K.; Cvelbar, U. Uniform surface growth of copper oxide nanowires in radiofrequency plasma discharge and limiting factors. *Phys. Plasmas (1994-Present)* **2014**, *21*.

31. Wang, A.; Qin, M.; Guan, J.; Wang, L.; Guo, H.; Li, X.; Wang, Y.; Prins, R.; Hu, Y. The synthesis of metal phosphides: Reduction of oxide precursors in a hydrogen plasma. *Angew. Chem.* **2008**, *120*, 6141–6143.

32. Vratnica, Z.; Vujosevic, D.; Cvelbar, U.; Mozetic, M. Degradation of bacteria by weakly ionized highly dissociated radio-frequency oxygen plasma. *IEEE Trans. Plasma Sci.* **2008**, *36*, 1300–1301.

33. Vesel, A.; Mozetic, M. Surface functionalization of organic materials by weakly ionized highly dissociated oxygen plasma. *J. Phys. Conf. Ser.* **2009**, *162*.

34. Hartney, M.A.; Greene, W.M.; Soane, D.S.; Hess, D.W. Mechanistic studies of oxygen plasma etching. *J. Vac. Sci. Technol. B* **1988**, *6*, 1892–1895.

35. Pearton, S.J.; Norton, D.P. Dry etching of electronic oxides, polymers, and semiconductors. *Plasma Process. Polym.* **2005**, *2*, 16–37.

36. Filipič, G.; Baranov, O.; Mozetič, M.; Cvelbar, U. Growth dynamics of copper oxide nanowires in plasma at low pressures. *J. Appl. Phys.* **2015**, *117*.

37. Cvelbar, U. Interaction of non-equilibrium oxygen plasma with sintered graphite. *Appl. Surf. Sci.***2013**, *269*, 33–36.

38. Cvelbar, U.; Mozetič, M.; Junkar, I.; Vesel, A.; Kovač, J.; Drenik, A.; Vrlinič, T.; Hauptman, N.; Klanjšek-Gunde, M.; Markoli, B.; *et al.* Oxygen plasma functionalization of poly(*p*-phenilene sulphide). *Appl. Surf. Sci.* **2007**, *253*, 8669–8673.

39. Wang, K.; Zhang, X.; Zhang, X.; Yang, B.; Li, Z.; Zhang, Q.; Huang, Z.; Wei, Y. Fabrication of cross-linked fluorescent polymer nanoparticles and their cell imaging applications. *J. Mater. Chem. C***2015**, *3*, 1854–1860.

40. Tsougeni, K.; Papageorgiou, D.; Tserepi, A.; Gogolides, E. "Smart" polymeric microfluidics fabricated by plasma processing: Controlled wetting, capillary filling and hydrophobic valving.*Lab Chip* **2010**, *10*, 462–469.

41. Yang, X.; Zhou, G.; Wong, W.-Y. Recent design tactics for high performance white polymer light-emitting diodes. *J. Mater. Chem. C* **2014**, *2*, 1760–1778.

42. Uysal Unalan, I.; Cerri, G.; Marcuzzo, E.; Cozzolino, C.A.; Farris, S. Nanocomposite films and coatings using inorganic nanobuilding blocks (NBB): Current applications and future opportunities in the food packaging sector. *RSC Adv.* **2014**, *4*, 29393–29428.

43. Vlachopoulou, M.-E.; Kokkoris, G.; Cardinaud, C.; Gogolides, E.; Tserepi, A. Plasma etching of poly(dimethylsiloxane): Roughness formation, mechanism, control, and application in the fabrication of microfluidic structures. *Plasma Process. Polym.* **2013**, *10*, 29–40.

44. Jacobs, T.; De Geyter, N.; Morent, R.; Desmet, T.; Dubruel, P.; Leys, C. Plasma treatment of polycaprolactone at medium pressure. *Surf. Coat. Technol.* **2011**, *205*, S543–S547.

45. Charbonnier, M.; Romand, M. Polymer pretreatments for enhanced adhesion of metals deposited by the electroless process. *Int. J. Adhes. Adhes.* **2003**, *23*, 277–285.

46. Yu, H.-D.; Regulacio, M.D.; Ye, E.; Han, M.-Y. Chemical routes to top-down nanofabrication.*Chem. Soc. Rev.* **2013**, *42*, 6006–6018.

47. Pochner, K.; Beil, S.; Horn, H.; Blömer, M. Treatment of polymers for subsequent metallization using intense UV radiation or plasma at atmospheric pressure. *Surf. Coat. Technol.* **1997**, *97*, 372–377.

48. Pascu, M.; Debarnot, D.; Durand, S.; Poncin-Epaillard, F. Surface modification of pvdf by microwave plasma treatment for electroless metallization. In *Plasma Processes and Polymers*; Wiley-VCH Verlag GmbH & Co. KGaA: Weinheim, Germany, 2005; pp. 157–176.

49. Macková, A.; Švorčík, V.; Strýhal, Z.; Pavlík, J. RBS and AFM study of Ag and Au diffusion into pet foils

influenced by plasma treatment. *Surf. Interface Anal.* **2006**, *38*, 335–338.

50. Lin, Y.S.; Liu, H.M. Enhanced adhesion of plasma-sputtered copper films on polyimide substrates by oxygen glow discharge for microelectronics. *Thin Solid Films* **2008**, *516*, 1773–1780.

51. Li, W.T.; Charters, R.B.; Luther-Davies, B.; Mar, L. Significant improvement of adhesion between gold thin films and a polymer. *Appl. Surf. Sci.* **2004**, *233*, 227–233.

52. Zille, A.; Fernandes, M.M.; Francesko, A.; Tzanov, T.; Fernandes, M.; Oliveira, F.R.; Almeida, L.; Amorim, T.; Carneiro, N.; Esteves, M.F.; *et al.* Size and aging effects on antimicrobial efficiency of silver nanoparticles coated on polyamide fabrics activated by atmospheric DBD plasma. *ACS Appl. Mater. Interfaces* **2015**, *7*, 13731–13744.

53. Boyd, R.D.; Kenwright, A.M.; Badyal, J.P.S.; Briggs, D. Atmospheric nonequilibrium plasma treatment of biaxially oriented polypropylene. *Macromolecules* **1997**, *30*, 5429–5436.

54. Bazaka, K.; Jacob, M.V.; Crawford, R.J.; Ivanova, E.P. Plasma-assisted surface modification of organic biopolymers to prevent bacterial attachment. *Acta Biomater.* **2011**, *7*, 2015–2028.

55. Poncin-Epaillard, F.; Brosse, J.C.; Falher, T. Cold plasma treatment: Surface or bulk modification of polymer films? *Macromolecules* **1997**, *30*, 4415–4420.

56. Yasuda, H.; Gazicki, M. Biomedical applications of plasma polymerization and plasma treatment of polymer surfaces. *Biomaterials* **1982**, *3*, 68–77.

57. Inglis, W.; Sanders, G.H.W.; Williams, P.M.; Davies, M.C.; Roberts, C.J.; Tendler, S.J.B. A simple method for biocompatible polymer based spatially controlled adsorption of blood plasma proteins to a surface. *Langmuir* **2001**, *17*, 7402–7405.

58. Wang, H.; Kwok, D.T.K.; Wang, W.; Wu, Z.; Tong, L.; Zhang, Y.; Chu, P.K. Osteoblast behavior on polytetrafluoroethylene modified by long pulse, high frequency oxygen plasma immersion ion implantation. *Biomaterials* **2010**, *31*, 413–419.

59. Oehr, C. Plasma surface modification of polymers for biomedical use. *Nucl. Instrum. Methods Phys. Res. Sect. B* **2003**, *208*, 40–47.

60. Halfmann, H.; Bibinov, N.; Wunderlich, J.; Awakowicz, P. A double inductively coupled plasma for sterilization of medical devices. *J. Phys. D* **2007**, *40*.

61. Vesel, A.; Kolar, M.; Stana-Kleinschek, K.; Mozetic, M. Etching rates of blood proteins, blood plasma and polymer in oxygen afterglow of microwave plasma. *Surf. Interface Anal.* **2014**, *46*, 1115–1118.

62. Stoffels, E.; Flikweert, A.J.; Stoffels, W.W.; Kroesen, G.M.W. Plasma needle: A non-destructive atmospheric plasma source for fine surface treatment of (bio)materials. *Plasma Sources Sci. Technol.* **2002**, *11*.

63. Herrmann, H.W.; Henins, I.; Park, J.; Selwyn, G.S. Decontamination of chemical and biological warfare (CBW) agents using an atmospheric pressure plasma jet (APPJ). *Phys. Plasmas* **1999**, *6*, 2284–2289.

64. Lazović, S.; Puač, N.; Miletić, M.; Pavlica, D.; Jovanović, M.; Bugarski, D.; Mojsilović, S.; Maletić, D.; Malović, G.; Milenković, P.; *et al.* The effect of a plasma needle on bacteria in planktonic samples and on peripheral blood mesenchymal stem cells. *New J. Phys.* **2010**, *12*.

65. Feng, L.; Li, S.; Li, Y.; Li, H.; Zhang, L.; Zhai, J.; Song, Y.; Liu, B.; Jiang, L.; Zhu, D. Super-hydrophobic surfaces: From natural to artificial. *Adv. Mater.* **2002**, *14*, 1857–1860.

66. Zhang, X.; Shi, F.; Niu, J.; Jiang, Y.; Wang, Z. Superhydrophobic surfaces: From structural control to functional application. *J. Mater. Chem.* **2008**, *18*, 621–633.

67. Erbil, H.Y.; Demirel, A.L.; Avcı, Y.; Mert, O. Transformation of a simple plastic into a superhydrophobic surface. *Science* **2003**, *299*, 1377–1380.

68. Puliyalil, H.; Filipič, G.; Cvelbar, U. Recent advances in the methods for designing superhydrophobic surfaces. In *Surface Energy*; Aliofkhazraei, M., Ed.; InTech: Rijeka, Croatia, 2015; pp. 311–335.

69. Toma, M.; Loget, G.; Corn, R.M. Flexible teflon nanocone array surfaces with tunable superhydrophobicity for self-cleaning and aqueous droplet patterning. *ACS Appl. Mater. Interfaces***2014**, *6*, 11110–11117.

70. Tsougeni, K.; Vourdas, N.; Tserepi, A.; Gogolides, E.; Cardinaud, C. Mechanisms of oxygen plasma nanotexturing of organic polymer surfaces: From stable super hydrophilic to super hydrophobic surfaces. *Langmuir* **2009**, *25*, 11748–11759.

71. Korshak, V.V.; Svetlana, V.V. Dependence of thermal stability of polymers on their chemical structure. *Russ. Chem. Rev.* **1968**, *37*, 885.

72. Taylor, G.N.; Wolf, T.M. Oxygen plasma removal of thin polymer films. *Polym. Eng. Sci.* **1980**, *20*, 1087–1092.

73. Moss, S.J.; Jolly, A.M.; Tighe, B.J. Plasma oxidation of polymers. *Plasma Chem. Plasma Process.* **1986**,*6*, 401–416.

74. Zadok, E.; Sialom, B.; Mazur, Y. Oxygen atoms produced by microwave discharge: Reaction with arenes. *Angew. Chem. Int. Ed. Engl.* **1980**, *19*, 1004–1005.

75. Zadok, E.; Rubinraut, S.; Frolow, F.; Mazur, Y. Reactions of di-, tri-, and hexamethylbenzenes with oxygen(3P) atoms in liquid and on adsorbed phases. *J. Am. Chem. Soc.* **1985**, *107*, 2489–2494.

76. Xie, L.; Jiao, L.; Dai, H. Selective etching of graphene edges by hydrogen plasma. *J. Am. Chem. Soc.***2010**, *132*, 14751–14753.

77. Xiang, H.; Kan, E.; Wei, S.-H.; Whangbo, M.-H.; Yang, J. “Narrow” graphene nanoribbons made easier by partial hydrogenation. *Nano Lett.* **2009**, *9*, 4025–4030.

78. Kerr, J.A. *Handbook of Bimolecular and Termolecular Gas Reactions*; Taylor & Francis: Boca Raton, FL, USA, 1987.

79. Mogab, C.J.; Adams, A.C.; Flamm, D.L. Plasma etching of Si and SiO_2—The effect of oxygen additions to CF_4 plasmas. *J. Appl. Phys.* **1978**, *49*, 3796–3803.

80. Chun, I.; Efremov, A.; Yeom, G.Y.; Kwon, K.-H. A comparative study of $CF_4/O_2/Ar$ and $C_4F_8/O_2/Ar$ plasmas for dry etching applications. *Thin Solid Films* **2015**, *579*, 136–143.

81. Hegemann, D.; Brunner, H.; Oehr, C. Plasma treatment of polymers for surface and adhesion improvement. *Nucl. Instrum. Methods Phys. Res. Sect. B* **2003**, *208*, 281–286.

82. Wohlfart, E.; Fernández-Blázquez, J.P.; Knoche, E.; Bello, A.; Pérez, E.; Arzt, E.; del Campo, A. Nanofibrillar patterns by plasma etching: The influence of polymer crystallinity and orientation in surface morphology. *Macromolecules* **2010**, *43*, 9908–9917.

83. Junkar, I.; Cvelbar, U.; Vesel, A.; Hauptman, N.; Mozetič, M. The role of crystallinity on polymer interaction with oxygen plasma. *Plasma Process. Polym.* **2009**, *6*, 667–675.

84. Chernomordik, B.D.; Russel, H.B.; Cvelbar, U.; Jasinski, J.B.; Kumar, V.; Deutsch, T.; Sunkara, M.K. Photoelectrochemical activity of as-grown, α-Fe_2O_3 nanowire array electrodes for water splitting.*Nanotechnology* **2012**, *23*.

85. Lazović, S.; Puač, N.; Spasić, K.; Malović, G.; Cvelbar, U.; Mozetič, M.; Radetić, M.; Petrović, Z.L. Plasma properties

in a large-volume, cylindrical and asymmetric radio-frequency capacitively coupled industrial-prototype reactor. *J. Phys. D Appl. Phys.* **2013**, *46*.

86. Olde Riekerink, M.B.; Terlingen, J.G.A.; Engbers, G.H.M.; Feijen, J. Selective etching of semicrystalline polymers: CF_4 gas plasma treatment of poly(ethylene). *Langmuir* **1999**, *15*, 4847–4856.

87. Nair, L.G.; Mahapatra, A.S.; Gomathi, N.; Joseph, K.; Neogi, S.; Nair, C.P.R. Radio frequency plasma mediated dry functionalization of multiwall carbon nanotube. *Appl. Surf. Sci.* **2015**, *340*, 64–71.

88. Black, C.T.; Guarini, K.W.; Milkove, K.R.; Baker, S.M.; Russell, T.P.; Tuominen, M.T. Integration of self-assembled diblock copolymers for semiconductor capacitor fabrication. *Appl. Phys. Lett.* **2001**,*79*, 409–411.

89. Chuang, V.P.; Ross, C.A.; Gwyther, J.; Manners, I. Self-assembled nanoscale ring arrays from a polystyrene-*b*-polyferrocenylsilane-*b*-poly(2-vinylpyridine)triblock terpolymer thin film. *Adv. Mater.* **2009**, *21*, 3789–3793.

90. Lammertink, R.G.H.; Hempenius, M.A.; van den Enk, J.E.; Chan, V.Z.H.; Thomas, E.L.; Vancso, G.J. Nanostructured thin films of organic–organometallic block copolymers: One-step lithography with poly(ferrocenylsilanes) by reactive ion etching. *Adv. Mater.* **2000**, *12*, 98–103.

91. Jung, Y.S.; Ross, C.A. Orientation-controlled self-assembled nanolithography using a polystyrene—Polydimethylsiloxane block copolymer. *Nano Lett.* **2007**, *7*, 2046–2050.

92. Kim, S.Y.; Nunns, A.; Gwyther, J.; Davis, R.L.; Manners, I.; Chaikin, P.M.; Register, R.A. Large-area nanosquare arrays from shear-aligned block copolymer thin films. *Nano Lett.* **2014**, *14*, 5698–5705.

93. Mozetič, M. Controlled oxidation of organic compounds in oxygen plasma. *Vacuum* **2003**, *71*, 237–240.

94. Panda, S.; Economou, D.J.; Meyyappan, M. Effect of metastable oxygen molecules in high density power-modulated oxygen discharges. *J. Appl. Phys.* **2000**, *87*, 8323–8333.

95. Dai, L.; Griesser, H.J.; Mau, A.W.H. Surface modification by plasma etching and plasma patterning. *J. Phys. Chem. B* **1997**, *101*, 9548–9554.

96. Chen, F.F.; Smith, M.D. Plasma. In *Van Nostrand's Scientific Encyclopedia*; John Wiley & Sons, Inc.: Hoboken, NJ, USA, 2005.

97. Donnelly, V.M.; Kornblit, A. Plasma etching: Yesterday, today, and tomorrow. *J. Vac. Sci. Technol. A***2013**, *31*.

98. Mozetič, M.; Ostrikov, K.; Ruzic, D.N.; Curreli, D.; Cvelbar, U.; Vesel, A.; Primc, G.; Leisch, M.; Jousten, K.; Malyshev, O.B.; *et al.* Recent advances in vacuum sciences and applications. *J. Phys. D***2014**, *47*.

99. Kontziampasis, D.; Constantoudis, V.; Gogolides, E. Plasma directed organization of nanodots on polymers: Effects of polymer type and etching time on morphology and order. *Plasma Process. Polym.* **2012**, *9*, 866–872.

100. Vourdas, N.; Kontziampasis, D.; Kokkoris, G.; Constantoudis, V.; Goodyear, A.; Tserepi, A.; Cooke, M.; Gogolides, E. Plasma directed assembly and organization: Bottom-up nanopatterning using top-down technology. *Nanotechnology* **2010**, *21*.

101. Puliyalil, H.; Cvelbar, U.; Filipic, G.; Petric, A.D.; Zaplotnik, R.; Recek, N.; Mozetic, M.; Thomas, S. Plasma as a tool for enhancing insulation properties of polymer composites. *RSC Adv.* **2015**, *5*, 37853–37858.

102. El-Aguizy, T.A.; Jeong, J.-H.; Jeon, Y.-B.; Li, W.Z.; Ren, Z.F.; Kim, S.-G. Transplanting carbon nanotubes. *Appl. Phys. Lett.* **2004**, *85*, 5995–5997.

103. Che, Y.; Wang, C.; Liu, J.; Liu, B.; Lin, X.; Parker, J.; Beasley, C.; Wong, H.S.P.; Zhou, C. Selective synthesis and device applications of semiconducting single-walled carbon nanotubes using isopropyl alcohol as feedstock. *ACS Nano* **2012**, *6*, 7454–7462.

104. Hou, P.-X.; Li, W.-S.; Zhao, S.-Y.; Li, G.-X.; Shi, C.; Liu, C.; Cheng, H.-M. Preparation of metallic single-wall carbon nanotubes by selective etching. *ACS Nano* **2014**, *8*, 7156–7162.

105. Zhang, G.; Qi, P.; Wang, X.; Lu, Y.; Li, X.; Tu, R.; Bangsaruntip, S.; Mann, D.; Zhang, L.; Dai, H. Selective etching of metallic carbon nanotubes by gas-phase reaction. *Science* **2006**, *314*, 974–977.

106. Di, C.-A.; Wei, D.; Yu, G.; Liu, Y.; Guo, Y.; Zhu, D. Patterned graphene as source/drain electrodes for bottom-contact organic field-effect transistors. *Adv. Mater.* **2008**, *20*, 3289–3293.

107. Liao, Z.; Wan, Q.; Liu, H.; Tang, Q. Realization of size controllable graphene micro/nanogap with a micro/nanowire mask method for organic field-effect transistors. *Appl. Phys. Lett.* **2011**, *99*.

108. Jiang, L.; Fan, Z. Design of advanced porous graphene materials: From graphene nanomesh to 3d architectures. *Nanoscale* **2014**, *6*, 1922–1945.

109. Song, X.; Hu, J.; Zeng, H. Two-dimensional semiconductors: Recent progress and future perspectives. *J. Mater. Chem. C* **2013**, *1*, 2952–2969.

110. Zang, J.; Ryu, S.; Pugno, N.; Wang, Q.; Tu, Q.; Buehler, M.J.; Zhao, X. Multifunctionality and control of the crumpling and unfolding of large-area graphene. *Nat. Mater.* **2013**, *12*, 321–325.

111. Zeng, Z.; Huang, X.; Yin, Z.; Li, H.; Chen, Y.; Li, H.; Zhang, Q.; Ma, J.; Boey, F.; Zhang, H. Fabrication of graphene nanomesh by using an anodic aluminum oxide membrane as a template.*Adv. Mater.* **2012**, *24*, 4138–4142.

112. Kim, M.; Safron, N.S.; Han, E.; Arnold, M.S.; Gopalan, P. Fabrication and characterization of large-area, semiconducting nanoperforated graphene materials. *Nano Lett.* **2010**, *10*, 1125–1131.

113. Ding, J.; Du, K.; Wathuthanthri, I.; Choi, C.-H.; Fisher, F.T.; Yang, E.-H. Transfer patterning of large-area graphene nanomesh via holographic lithography and plasma etching. *J. Vac. Sci. Technol. Technol. B* **2014**, *32*.

114. Sinitskii, A.; Tour, J.M. Patterning graphene through the self-assembled templates: Toward periodic two-dimensional graphene nanostructures with semiconductor properties. *J. Am. Chem. Soc.* **2010**, *132*, 14730–14732.

115. Liang, X.; Jung, Y.-S.; Wu, S.; Ismach, A.; Olynick, D.L.; Cabrini, S.; Bokor, J. Formation of bandgap and subbands in graphene nanomeshes with sub-10 nm ribbon width fabricated via nanoimprint lithography. *Nano Lett.* **2010**, *10*, 2454–2460.

116. Paul, R.K.; Badhulika, S.; Saucedo, N.M.; Mulchandani, A. Graphene nanomesh as highly sensitive chemiresistor gas sensor. *Anal. Chem.* **2012**, *84*, 8171–8178.

117. Fatyeyeva, K.; Dahi, A.; Chappey, C.; Langevin, D.; Valleton, J.-M.; Poncin-Epaillard, F.; Marais, S. Effect of cold plasma treatment on surface properties and gas permeability of polyimide films. *RSC Adv.* **2014**, *4*, 31036–31046.

118. Zarshenas, K.; Raisi, A.; Aroujalian, A. Surface modification of polyamide composite membranes by corona air plasma for gas separation applications. *RSC Adv.* **2015**, *5*, 19760–19772.

119. Chen, S.-H.; Chuang, W.-H.; Wang, A.A.; Ruaan, R.-C.; Lai, J.-Y. Oxygen/nitrogen separation by plasma chlorinated polybutadiene/polycarbonate composite membrane. *J. Membr. Sci.* **1997**, *124*, 273–281.

120. Chen, Z.; Tan, X. Monolithic fabrication of ionic polymer–metal composite actuators capable of complex deformation. *Sens. Actuators A* **2010**, *157*, 246–257.

121. Nikhil Dilip, P.; Mark, S.M. An investigation of the fracturing process in nitrogen-doped multiwalled carbon nanotubes (N-MWCNTs). Evidence for directional unzipping. *Mater. Res. Express* **2014**, *1*.

122. Jiao, L.; Zhang, L.; Wang, X.; Diankov, G.; Dai, H. Narrow graphene nanoribbons from carbon nanotubes. *Nature* **2009**, *458*, 877–880.

123. Poulsen, R.G. Plasma etching in integrated circuit manufacture—A review. *J. Vac. Sci. Technol.* **1977**, *14*, 266–274.

124. Vitale, S.A.; Berry, S. Etching selectivity of indium tin oxide to photoresist in high density chlorine- and ethylene-containing plasmas. *J. Vac. Sci. Technol. B* **2013**, *31*.

125. Laermer, F.; Urban, A. Challenges, developments and applications of silicon deep reactive ion etching. *Microelectron. Eng.* **2003**, *67–68*, 349–355.

126. Liao, Q.; Xu, Z.; Zhong, X.; Dang, W.; Shi, Q.; Zhang, C.; Weng, Y.; Li, Z.; Fu, H. An organic nanowire waveguide exciton-polariton sub-microlaser and its photonic application. *J. Mater. Chem. C* **2014**, *2*, 2773–2778.

127. Garcia-Frutos, E.M. Small organic single-crystalline one-dimensional micro- and nanostructures for miniaturized devices. *J. Mater. Chem. C* **2013**, *1*, 3633–3645.

128. Fang, H.; Wu, W.; Song, J.; Wang, Z.L. Controlled growth of aligned polymer nanowires. *J. Phys. Chem. C* **2009**, *113*, 16571–16574.

129. Chen, Y.; Xu, Z.; Gartia, M.R.; Whitlock, D.; Lian, Y.; Liu, G.L. Ultrahigh throughput silicon nanomanufacturing by simultaneous reactive ion synthesis and etching. *ACS Nano* **2011**, *5*, 8002–8012.

130. Chandran, N.; Chandran, S.; Maria, H.J.; Thomas, S. Compatibilizing action and localization of clay in a polypropylene/natural rubber (PP/NR) blend. *RSC Adv.* **2015**, *5*, 86265–86273.

131. Jyotishkumar, P.; Koetz, J.; Tiersch, B.; Strehmel, V.; Özdilek, C.; Moldenaers, P.; Hässler, R.; Thomas, S. Complex phase separation in poly(acrylonitrile–butadiene–styrene)-modified epoxy/4,4′-diaminodiphenyl sulfone blends: Generation of new micro- and nanosubstructures. *J. Phys. Chem. B* **2009**, *113*, 5418–5430.

132. Kunaver, M.; Mozetič, M.; Klanjšek-Gunde, M. Selective plasma etching of powder coatings. *Thin Solid Films* **2004**, *459*, 115–117.

133. Kunaver, M.; Klanjsek-Gunde, M.; Mozetic, M.; Kunaver, M.; Hrovat, A. The degree of dispersion of pigments in powder coatings and the origin of some surface defects. *Surf. Coat. Int. Part B* **2003**,*86*, 175–179.

134. Mozetič, M.; Zalar, A.; Panjan, P.; Bele, M.; Pejovnik, S.; Grmek, R. A method of studying carbon particle distribution in paint films. *Thin Solid Films* **2000**, *376*, 5–8.

135. Salalha, W.; Dror, Y.; Khalfin, R.L.; Cohen, Y.; Yarin, A.L.; Zussman, E. Single-walled carbon nanotubes embedded

in oriented polymeric nanofibers by electrospinning. *Langmuir* **2004**, *20*, 9852–9855.

136. Ponnamma, D.; Sadasivuni, K.K.; Grohens, Y.; Guo, Q.; Thomas, S. Carbon nanotube based elastomer composites—An approach towards multifunctional materials. *J. Mater. Chem. C* **2014**, *2*, 8446–8485.

137. Cvelbar, U.; Pejovnik, S.; Mozetiè, M.; Zalar, A. Increased surface roughness by oxygen plasma treatment of graphite/polymer composite. *Appl. Surf. Sci.* **2003**, *210*, 255–261.

138. Malara, F.; Manca, M.; De Marco, L.; Pareo, P.; Gigli, G. Flexible carbon nanotube-based composite plates as efficient monolithic counter electrodes for dye solar cells. *ACS Appl. Mater. Interfaces* **2011**, *3*, 3625–3632.

139. Zhang, Y.; Li, J.; An, G.; He, X. Highly porous SnO_2 fibers by electrospinning and oxygen plasma etching and its ethanol-sensing properties. *Sens. Actuators B* **2010**, *144*, 43–48.

140. Ramgir, N.; Datta, N.; Kaur, M.; Kailasaganapathi, S.; Debnath, A.K.; Aswal, D.K.; Gupta, S.K. Metal oxide nanowires for chemiresistive gas sensors: Issues, challenges and prospects. *Colloids Surf. A* **2013**, *439*, 101–116.

141. Bandgar, D.K.; Navale, S.T.; Nalage, S.R.; Mane, R.S.; Stadler, F.J.; Aswal, D.K.; Gupta, S.K.; Patil, V.B. Simple and low-temperature polyaniline-based flexible ammonia sensor: A step towards laboratory synthesis to economical device design. *J. Mater. Chem. C* **2015**, *3*, 9461–9468.

142. Han, J.-W.; Kim, B.; Li, J.; Meyyappan, M. Carbon nanotube based humidity sensor on cellulose paper. *J. Phys. Chem. C* **2012**, *116*, 22094–22097.

143. Raghu, M.; Suresh, R.; Vijay, P.S. MWCNT–polymer composites as highly sensitive and selective room temperature gas sensors. *Nanotechnology* **2011**, *22*.

144. Ernest, M.; Jahir, O.; Cecilia, J.-J.; Ana, B.G.-G.; Ana, C.; Laura, M.L.; César, F.-S. Scalable fabrication of immunosensors based on carbon nanotube polymer composites. *Nanotechnology***2008**, *19*.

145. Sullalti, S.; Colonna, M.; Berti, C.; Fiorini, M.; Karanam, S. Effect of phosphorus based flame retardants on UL94 and comparative tracking index properties of poly(butylene terephthalate).*Polym. Degrad. Stab.* **2012**, *97*, 566–572.

3
CHAPTER

Electrochemistry-Based Maskless Nanofabrication

I. Tiginyanu

Lab. of Nanotechnol., Acad. of Sci. of Moldova, Chisinau, Moldova

ABSTRACT

A review of technological approaches for 2D and 3D nanostructuring of semiconductor compounds by using radiation treatment and electrochemical etching is presented. We demonstrate novel spatial nanoarchitectures based on III-V and II-VI compounds as well as two-dimensional metallo-dielectric structures realized in different geometries. It is shown that photoelectrochemical etching of GaN combined with preliminary low-dose low-energy focused-ion-beam treatment of the sample surface allows one to fabricate in a controlled fashion arrays of nanowires and nanowalls as well as ultrathin membranes and supporting nanocolumns in the same technological route. Possible electronic and photonic applications of the elaborated nanostructures are discussed.

Keywords: semiconductors, Nanostructuring, etching, electroplating, nanowires, ultrathin membranes

INTRODUCTION

The technologies allowing one to manipulate with the spatial architecture of materials at the nanometer scale become more and more expensive when they are related to nano-lithographic top-down approaches or to precise handling with each atom or molecule used as building blocks in bottom-up approaches. Over the last decade, considerable research efforts were undertaken to develop cost-effective top-down and bottom-up nanotechnologies based on self-organization phenomena and self-assembling. In this review paper, we present new developments in top-down non-lithographic nanotechnologies based on electrochemical or photoelectrochemical (PEC) etching. We report, in particular, on maskless fabrication of cost-effective III - V and II - VI semiconductor nanotemplates, 2D quasi-periodic metallo-semiconductor structures and 3D spatial nanoarchitectures consisting of ultrathin membranes and supporting nanocolumns or nanowalls, the design of both membranes and nanocolumns/nanowalls being realized by focused ion beam (FIB) direct writing.

CONDUCTIVE NANOTEMPLATES

Two types of nanotemplates have been developed and widely used over the last decade for nanofabrication purposes, namely porous aiumina (Al2O3) and etched ion track membranes based either on inorganic materials or on organic polymers [1][2][3][4]. These types of porous membranes, however, do not possess electrical conductivity and therefore they play only a passive role in nanofabrication processes. In this connection one of the goals of our efforts was to develop multifunctional semiconductor nanotemplates.

We found that anodic etching of single crystalline GaP, InP, ZnSe and CdSe in acidic solutions such as HCI, H2SO4,HNO3 etc. leads to material porosification, although the porous architecture develops in different ways [5][6][7][8][9]. In GaP and InP compounds, porosification starts with the formation of crystallographically oriented pores in the so called nucleation layer. After multiple branching of the crystallographically oriented pores, anodic etching starts to produce current-line oriented pores exhibiting a pronounced tendency to form rows oriented along < 110> direction. This tendency accompanied by the repulsive pore-pore interaction due to overlapping surface depletion layers surrounding neighboring pores leads to self-arrangement of pores and their ordered close packed 2D distribution.

No crystallographically oriented pores have been observed in II-VI compounds so far. After nucleation at the surface defects, the pores prove to be oriented along the current lines, exhibiting multiplication until the front of the porous network covers the whole available space [10][11]. Along with the uniform porosification, in ZnSe we demonstrated the formation of multilayer porous structures, including layers subjected to successive porosification at two different length scales [11]. Besides, growth of fractal-like pores has been realized in ZnSe single crystals for the first time, seeFig. 1.

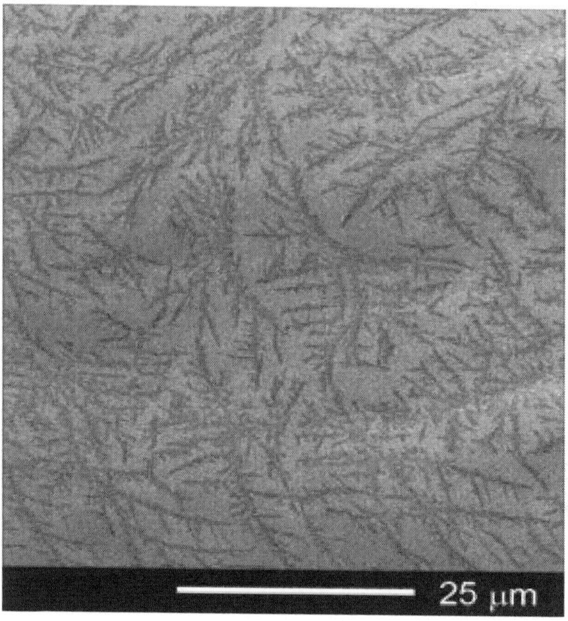

Fig. 1. Fractal pores in single crystalline ZnSe.

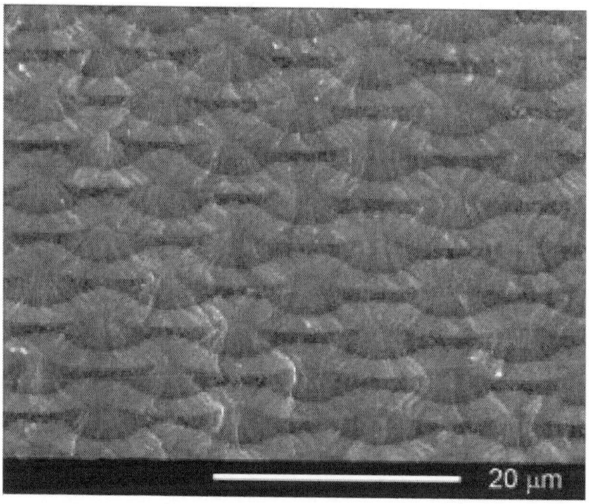

Fig. 2. Scanning Electron Microscope (SEM) image taken from a porous InP sample anodized under periodic modulation of the applied voltage with time.

Interestingly, formation of quasi-ordered two-dimensional hexagonal arrays of pores in some III - V and II -VI compounds can be realized in environmentally friendly electrolytes like aqueous solution of NaCl [12][13]. We demonstrated that, depending upon anodization conditions, both 2D and 3D nanostructuring can be achieved in single crystalline InP [12]. Figure 2 illustrates a 3D periodic structure on n-InP fabricated by anodic etching carried out under quasi-potentiostatic conditions, the value of the applied voltage being periodically modulated with time.

Semiconductor nanotemplates possess electrical conductivity and photoconductivity [14][15], their values depending upon the characteristics of the huge surface inherent to porous structures. Recently we found that the surface charge density of nanoporous InP can be efficiently altered by photoexcitation [16]. Besides, nanotemplates based on 111- V compounds exhibit artificial birefringence [17][18], enhanced optical second harmonic generation [18][19], and enhanced Terahertz emission [19][20]. According to the results of our exploration [21], irradiating the nanoporous InP membranes with heavy noble gas Kr or Xe ions leads to considerable intensification of the terahertz emission under near-infrared optical excitation. Systematic investigation of the dependence of the generated terahertz electric field on excitation pump power, in-plane magnetic field, and azimuthal angle indicates that the underlying physical mechanism is optical rectification rather than transient current flow [21]. These data are important for further improvement in solid-state terahertz emitters.

TWO DIMENSIONAL METALLO-SEMICONDUCTOR STRUCTURES

Over the last decade, considerable research interest was paid to one dimensional metallo-dielectric multilayer structures which prove to be transparent in specific spectral regions and exhibit negative refraction and subwavelength imaging [22][23]. Recently we proved analytically that metallized titania nanotubes are promising for designing and manufacturing negative refractive index materials [24]. At the same time we have undertaken technological efforts to develop two dimensional metallo-semiconductor structures by using electrochemical deposition of metals in semiconductor nanotemplates [6][25][26]. Applying electroplating of Pt from a commercial platinum bath under pulsed voltage regime, we succeeded to fabricate ordered and quasi-ordered arrays of Pt nanotubes embedded in porous matrices of GaP, InP and ZnSe. Note that in this case the semiconductor nanotemplates play an active role in the process of electrochemical deposition. We found that electroplating starts with the deposition on the inner surface of pores of metal dots, their density increasing with time. Overlapping of the neighboring dots leads to the formation of metal nanotubes exhibiting rather smooth surfaces [6][25][26]. Fig. 3(a) presents a SEM image taken from an n-GaP nanotemplate with the pores self-arranged in a two-dimensional hexagonal array, while Fig. 3(b) illustrates an array of Pt nanotubes embedded in the porous GaP matrix. Note that on the SEM image the metal nanotubes look bright when they get out from the semiconductor matrix. As mentioned in Ref. 25, this is a consequence of the charging phenomenon caused by the potential barrier at the Pt-InP interface. The formation of a Schottky barrier between the array of Pt nanotubes and semiconductor matrix allowed one to develop a variable capacitance device with enhanced capacitance density variation for a small change in voltage [26].

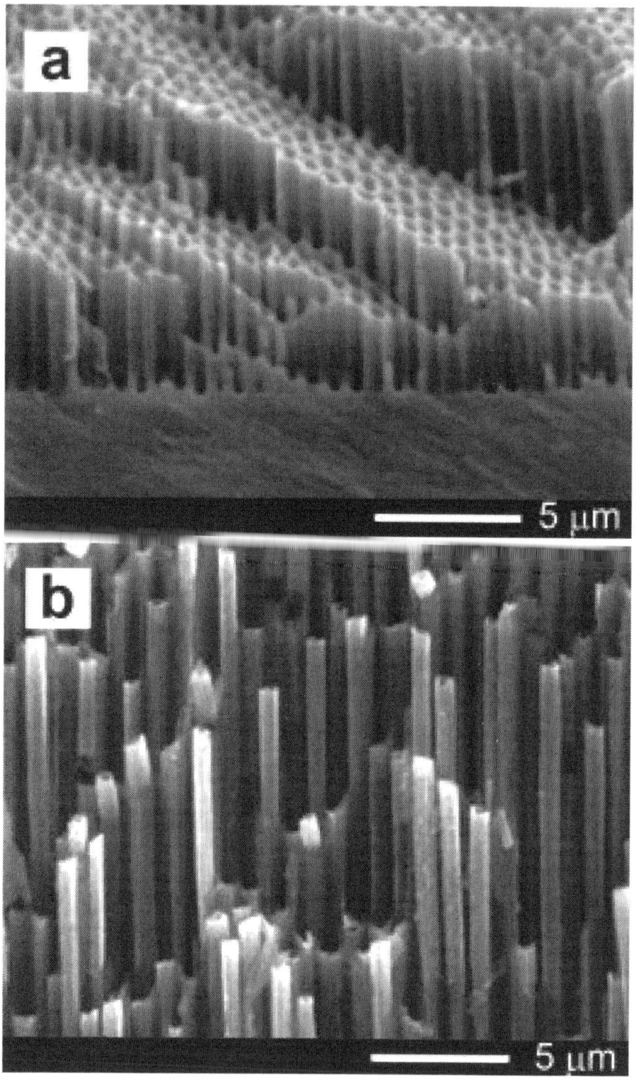

Fig. 3. SEM images taken from a GaP nanotemplate before (a) and after (b) pulsed electroplating of Pt. The samples have been cleaved to explore the morphology in depth.

SURFACE CHARGE LITHOGRAPHY

Previously we proposed the approach of surface charge lithography (SCL) as a tool for maskless micro- and nanostructuring of GaN [27][28][29]. This approach is based on treatment of the semiconductor surface by a low-dose low-energy ion beam with subsequent photoelectrochemical etching of the sample. The ion-beam induced lattice defects trap electrons leading to the appearance of a surplus of negative charge in the near-surface region of the GaN sample. It is the negative charge that protects the ion-beam treated areas against PEC etching. Using the ion-beam-induced negative charge as a shield against PEC etching, we demonstrate unique possibilities for GaN nanostructuring, including fabrication of ultrathin membranes suspended over nanowalls fabricated in the same technological route.

Fig. 4 illustrates GaN mesastructures fabricated by using Ar^+ ion treatment of selected surface areas with subsequent PEC etching of the GaN epilayer. The negative charge trapped by the implantation-induced defects plays the role of lithographic mask during PEC etching. The proposed maskless techniques allow one to fabricate also nanowire-like structures in a controlled fashion. Fig. 5 illustrates nanowire fabrication by 30-keV Ga+ FIB direct "writing" of the negative charge with subsequent PEC etching under controlled conditions. Note that the nanowire merges GaN mesastructures which can be used as platforms for the fabrication of electrical contacts.

Fig. 4. SEM image taken from a GaN layer subjected to PEC etching after selected areas were treated by Ar^+ ions (E=2keV, 3×1012 cm−2). The letters “survived” during PEC etching due to the presence of the shielding negative charge.

To fabricate ultrathin suspended membranes we used the same technological route consisting of low-energy/low-dose Ar^+ or Ga^+ ion treatment and PEC etching.

Being defined by the ion fluence, the density of implantation-induced defects is in this case as low as to provide good transparency to UV radiation of the emerging membrane. Under such conditions the bulk of the GaN epilayer is photoelectrochemically etched, with the exception of the ultrathin negatively-charged surface film and of the so-called whiskers representing threading dislocations which prove to be negatively-charged as well [30][31][32]. Fig. 6 illustrates a nanometer-thick GaN membrane hanging over a network of whiskers with the diameters of about 50 nm.

Fig. 5. SEM image taken from GaN layer subjected to PEC etching after selected areas were treated by Ga^{+}(E=30keV,6.6×1012cm−2) focused ion beam.

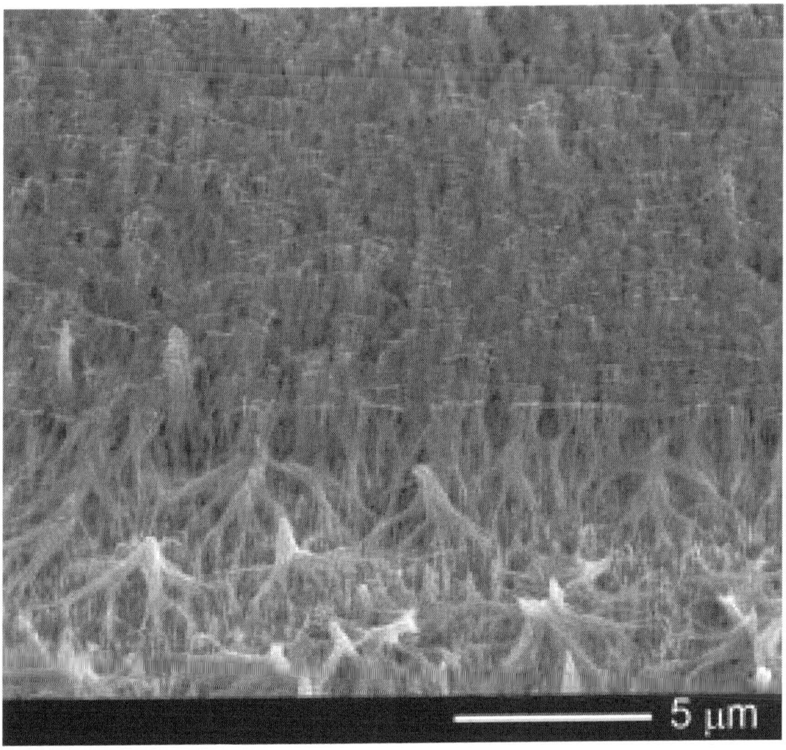

Fig. 6. SEM image taken from a GaN membrane hanging over a network of whiskers representing threading dislocations. The fluence of Ar^{+} ions is 10^{11} cm^{-2}.

It is to be noted that the formation of nanometer-thick suspended membrane to which the dislocations are genetically attached provides conditions for the revelation in GaN of the spatial nanoarchitecture of dislocation networks using conventional SEM, see Ref. 31 for details.

ULTRATHIN MEMBRANES FOR MEMS/NEMS APPLICATIONS

Recently we found technological conditions for removal of whiskers during PEC etching. Further, we identified an elegant way of fabrication in the same technological route of both ultrathin GaN membranes and supporting micro/nanocolumns (or micro/nanowalls) [33]. The essence of the approach is as follows. A rectangular area of GaN layer is subjected to low-energy/low-dose ion beam treatment which provides conditions for the formation of an ultrathin membrane under subsequent PEC etching. To avoid stiction of the membrane to the bottom surface, some selected areas are subjected to high-dose ion implantation which creates prerequisites for the formation during PEC etching of micro/nanocolumns supporting the emerging membrane. Note that the high-dose ion-treated top areas efficiently absorb UV radiation due to the high density of implantation-induced defects.

Fig. 7 illustrates a SEM image taken from a rectangular 15-nm thick membrane suspended over specially designed columns and walls. The supporting structures have been "written" by the high-dose 30-keV Ga^+ focused ion beam.

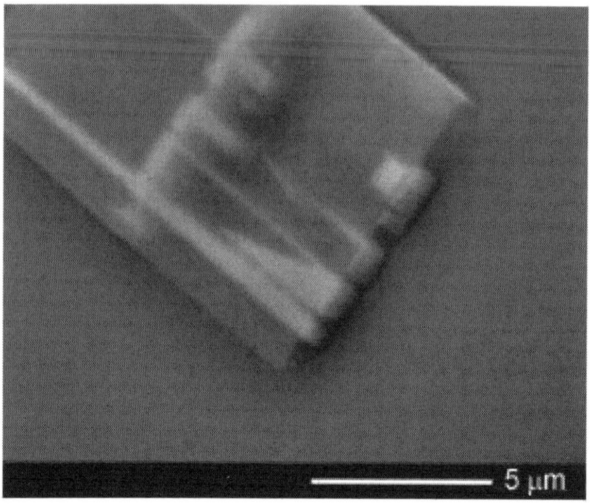

Fig. 7. SEM image (top oblique view) taken from a GaN membrane suspended over micro/nanocolumns and micro/nanowalls "written" by the Ga + focused ion beam at a relatively high fluence [33].

According to recently published results [30][31][32], the nanometer-thick GaN membranes exhibit luminescence, the spectrum being prevailed by yellow emission when excited by electron beams. Comparing our results with the data obtained by Li and Wang [34], one can claim that the near-surface few-atoms-thick regions of GaN (at least in nanostructures) emit mainly yellow luminescence. Preliminary exploration of the conductivity of GaN nanostructures evidenced an interesting behavior. Although 100-nm thick GaN nanowalls are non-conductive due to carrier depletion, nanometer-thick GaN membranes prove to exhibit rather good electrical conductivity and photoconductivity. Besides, after excitation by UV radiation at relatively low temperatures, the membranes involved are characterized by pronounced persistent photoconductivity.

It is interesting to note that, over the last years, many researchers have focused their efforts at manufacturing ultrathin solid membranes, including sheets of few-atomic layer graphene, boron nitride (BN), molybdenum disulfide (MoS_2), bismuth telluride (Bi_2Te_3), bismuth selenide (Bi_2Se_3), and so on. To our knowledge, however, with the exception of our recent paper [33], all reported crystalline sheets are based on layered materials characterized by strongly anisotropic chemical bonding, where adjacent structural units are coupled by weak van der Waals interaction. It is this weak interaction that allows mechanical exfoliation of ultrathin membranes. Successfully fabricated nanometerthin GaN membranes open unique opportunities for exploration of the properties of 2D sheets based on non-layered crystalline solids.

CONCLUSIONS

Electrochemistry proves to be a powerful tool for nanofabrication. Using anodic etching techniques, we demonstrated the possibility to produce semiconductor nanotemplates with ordered or quasi-ordered distribution of pores in two dimensional hexagonal arrays. The relatively good electrical conductivity of the porous skeleton allowed us to uniformly deposit Pt on the inner surface of pores by electroplating, leading to the formation of arrays of metal nanotubes embedded in the semiconductor matrix. Additional attention in the near future should be paid to electrochemical deposition of metals in porous GaAs exhibiting only crystallographically oriented pores growing along <111> directions [35][36] or, in other words, exhibiting three dimensional pore lattices promising for various photonic applications. An important result is the evidenced possibility to design and fabricate both ultrathin membranes and supporting micro/nanocolumns (or walls) in the same maskless technological route. Taking into account the piezoelectric properties inherent to GaN, designable fabrication of ultrathin suspended GaN membranes provides unique possibilities for various MEMS/NEMS applications.

ACKNOWLEDGEMENTS

The authors wish to acknowledge financial support from the European Commission under the FP7 project MOLD-ERA (Grant no 266515), and from the Supreme Council for Research and Technological Development (Academy of Sciences of Moldova) under the State Program on Nanotechnologies and Nanomaterials.

REFERENCES

1. I.U. Schuchert, M.E. Toimil Molares, D. Dobrev, J. Vetter, R. Neumann and M. Martin, "Electrochemical copper deposition in etched ion track membranes", *J. Electrochem. Soc.*, vol. 150, pp. 189-194, 2003.

2. W. Lee, R. Scholz, K. Nielsch and U. Gösele, "A template-based electrochemical method for the synthesis of multisegmented metallic nanotubes", *Angew. Chem. Int. Ed.*, vol. 44, pp. 650-654, 2005.

3. H. Koshikawa, Y. Maekawa and H. Usui, "Formation of composite films of ion-track membranes embedded with oblique Cu nanowires for anisotropic infrared absorption", *Rad. Phys. Chem.*, vol. 77, no. 4, pp. 453-455, 2008.

4. C.T. Sousa, D.C. Leitao, M.P. Proenca, A. Apolinario, J.G. Correia, J. Ventura, et al., "Tunning pore filling of anodic alumina templates by accurate control of the bottom barrier layer thickness", *Nanotechnology*, vol. 22, pp. 315602, 2011.

5. S. Langa, I.M. Tiginyanu, E. Mosaico and H. Föll, "Porous II-VI vs. porous III-V semiconductors", *Phys. Stat. Sol. C*, vol. 8, no. 6, pp. 1792-1796, 2011.

6. I.M. Tiginyanu, V.V. Ursaki, E. Monaico, M. Enachi, V.V. Sergentu, G. Colibaba, et al., "Quasi-ordered networks of metal nanotubes embedded in semiconductor matrices for photonic applications", *J. Nanoelectron. Optoelectron.*, vol. 6, no. 4, pp. 463-472, 2011.

7. I.M. Tiginyanu, E. Monaico, V.V. Ursaki, V.E. Tezlevan and R.W. Boyd, "Fabrication and photoluminescence properties of porous CdSe", *Appl. Phys. Lett.*, vol. 86, pp. 063115, 2005.

8. S. Langa, J. Carstensen, M. Christophersen, K. Steen, S. Frey, I.M. Tiginyanu, et al., "Uniform and nonuniform nucleation of pores during the anodization of Si Ge and III-V semiconductors", *J. Electrochem. Soc.*, vol. 152, no. 8, 2005.

9. S. Langa, M. Christophersen, J. Carstensen, I.M. Tiginyanu and H. Föll, "Single crystalline 2D porous arrays obtained

by self organization in n-InP", *Phys. Stat. Sol. A*, vol. 197, no. 1, pp. 77-82, 2003.

10. E. Monaico, V.V. Ursaki, I.M. Tiginyanu, Z. Dashevsky, V. Kasiyan and R.W. Boyd, "Porosity-induced blueshift of photoluminescence in CdSe", *J. Appl. Phys.*, vol. 100, pp. 053517, 2006.

11. E. Monaico, P. Tighineanu, S. Langa, H.L. Hartnagel and I. Tiginyanu, "ZnSe-based conductive nanotemplates for nanofabrication", *Phys. Stat. Sol. RRL*, vol. 3, no. 4, pp. 97-99, 2009.

12. I.M. Tiginyanu, E. Monaico, S. Albu and V.V. Ursaki, "Environmentally friendly approach for nonlithographic nanostructuring of materials", *Phys. Stat. Sol. RRL*, vol. 1, no. 3, pp. 98-100, 2007.

13. I.M. Tiginyanu, E. Monaico, V.V. Ursaki, E. Foca and H. Föll, "Pore etching in III-V and II-VI semiconductor compounds in neutral electrolyte", *Electrochem. Sol.-St. Lett.*, vol. 10, no. 11, 2007.

14. S.K.E. Merchant, J. Lloyd-Hughes, L. Sirbu, I.M. Tiginyanu, P. Parkinson, L.M. Herz, et al., "Conductivity of nanoporous InP membranes investigated using terahertz spectroscopy",*Nanotechnology*, vol. 19, pp. 395704, 2008.

15. J. Lloyd-Hughes, S.K.E. Merchant, L. Sirbu, I.M. Tiginyanu and M.B. Johnston, "Terahertz photoconductivity of mobile electrons in nanoporous InP honeycombs", *Phys. Rev. B*, vol. 78, pp. 085320, 2008.

16. J. Lloyd-Hughes, S. Müller, G. Scalari, H. Bishop, A. Crossley, M. Enachi, et al., "Photoinduced modification of surface

states in nanoporous InP", *Appl. Phys. Lett.*, vol. 100, no. 2012, pp. 132106.

17. V.V. Ursaki, N.N. Syrbu, S. Albu, V.V. Zalamai, I.M. Tiginyanu and R.W. Boyd, "Artificial birefringence introduced by porosity in GaP", *Semicond. Sci. Technol.*, vol. 20, pp. 745-748, 2005.

18. I.M. Tiginyanu, I.V. Kravetsky, S. Langa, G. Marowsky, J. Monecke and H. Föll, "Porous III-V compounds as nonlinear optical materials", *Phys. Stat. Sol. A*, vol. 197, no. 2, pp. 549-555, 2003.

19. M. Reid, I.I.V. Cravetchi, R. Fedosejevs, I.M. Tiginyanu, L. Sirbu and R.W. Boyd, "Enhanced nonlinear optical response of InP(100) membranes", *Phys. Rev. B*, vol. 71, pp. 081306, 2005.

20. M. Reid, I.V. Cravetchi, R. Fedosejevs, I.M. Tiginyanu and L. Sirbu, "Enhanced terahertz emission from porous InP(111) membranes", *Appl. Phys. Lett.*, vol. 86, pp. 021904, 2005.

21. K. Radhanpura, S. Hargreaves, R.A. Lewis, L. Sirbu and I.M. Tiginyanu, "Heavy noble gas (Kr Xe) irradiated (111) InP nanoporous honeycomb membranes with superior optical nonlinearity and enhanced ultrafast all-optical Terahertz emission", *Appl. Phys. Lett.*, vol. 97, pp. 181921, 2010.

22. M. Scalora, G. D'Aguanno, N. Mattiucci and M.J. Bloemer, "Negative refraction and sub-wavelength focusing in the visible range using transparent metallo-dielectric stacks", *Opt. Express*, vol. 15, no. 2, pp. 508-523, 2007.

23. M. Bloemer, G. D'Aguanno, M. Scalora, N. Mattiucci and D. De Ceglia, "Energy considerations for a superlens based on metal/dielectric multilayers", *Opt. Express*, vol. 16, no. 23, pp. 19342-19353, 2008.

24. V.V. Sergentu, I.M. Tiginyanu, V.V. Ursaki and M. Enachi, "Prediction of negative index material lenses based on metallo-dielectric nanotubes", *Phys. Stat. Sol. RRL*, vol. 2, pp. 242-244, 2008.

25. Tiginyanu, E. Monaico and E. Monaico, "Ordered arrays of metal nanotubes in semiconductor envelope", *Electrochem. Commun.*, vol. 10, pp. 731-734, 2008.

26. Tiginyanu, E. Monaico and V. Ursaki, "Two-Dimensional Metallo-Semiconductor Networks for Electronic and Photonic Applications", *ECS Transactions*, vol. 41, no. 44, pp. 67-74, 2012.

27. I.M. Tiginyanu, V. Popa and O. Volciuc, "Surface-charge lithography for GaN microstructuring based on photoelectrochemical etching techniques", *Appl. Phys. Lett.*, vol. 86, pp. 174102, 2005.

28. V. Popa, I. Tiginyanu, O. Volciuc, A. Sarua, M. Kuball and P. Heard, "Fabrication of GaN nanowalls and nanowires using surface charge lithography", *Mater. Lett.*, vol. 62, pp. 4576-4578, 2008.

29. I.M. Tiginyanu, V. Popa, A. Sarua, P.J. Heard, O. Volciuc and M. Kuball, "Surface charge lithography for GaN micro- and nanostructuring", *Proc. SPIE*, vol. 7216, 2009.

30. I.M. Tiginyanu, V. Popa and M.A. Stevens-Kalceff, "Nanoperforated and Continuous Ultra-Thin GaN Membranes", *Electrochem. Sol.-St. Lett.*, vol. 14, no. 9, 2011.

31. I.M. Tiginyanu, V. Popa and M.A. Stevens-Kalceff, "Membrane-assisted revelation of the spatial nanoarchitecture of dislocation networks", *Mater. Lett.*, vol. 65, no. 2, pp. 360-362, 2011.

32. Tiginyanu, V. Popa and M.A. Stevens-Kalceff, "Ultra-thin semiconductor membrane nanotechnology based on surface charge lithography", *Proc. SPIE*, vol. 8068, pp. 806814, 2011.

33. I.M. Tiginyanu, V. Popa, M.A. Stevens-Kalceff, D. Gerthsen, P. Brenner and D. Pavlidis, "Design and maskless fabrication of ultrathin suspended membranes of GaN", *Phys. Stat. Sol. RRL*, vol. 6, no. 4, pp. 148-150, 2012.

34. Q. Li and G.T. Wang, "Spatial distribution of defect luminescence in GaN nanowires", *Nano Lett.*, vol. 10, pp. 1554-1558, 2010.

35. H. Föll, S. Langa, J. Carstensen, M. Christophersen and I.M. Tiginyanu, "Pores in III-V semiconductors", *Adv. Mater.*, vol. 15, no. 3, pp. 183-198, 2003.

36. S. Langa, J. Carstensen, M. Christophersen, H. Föll and I.M. Tiginyanu, "Observation of crossing pores in anodically-etched n-GaAs", *Appl. Phys. Lett.*, vol. 78, no. 8, pp. 1074-1076, 2001.

4

CHAPTER

Nanobowl Array Fabrication via Conglutination Process Based on Thiol–Ene Polymer

Man Zhang,[1] Liangping Xia,[2] Qiling Deng,[1] Lifang Shi,[1] Axiu Cao,[1] Hui Pang,[1] and Song Hu

[1]*Institute of Optics and Electronics, Chinese Academy of Sciences, Chengdu 610209, China*

[2] *Chongqing Institute of Green and Intelligent Technology, Chinese Academy of Sciences, Chongqing 401122, China*

ABSTRACT

The technique to fabricate nanostructures with high resolution is of crucial importance to nanosciences and nanotechnology. This paper presented a nonconventional method to fabricate metallic nanobowl arrays via the conglutination process, which relied on the high adhesive material to "stick" nanostructures. The ultraviolet (UV)-curable thiol-ene had low viscosity and was cured under UV light to form a cross-linked polymer with high

Young's modulus and high surface energy via "click chemistry," which was used as the adhesive material to firmly stick nanostructures from the mold. By using the method, a self-assembly polystyrene (PS) sphere template with a metallic layer was stuck by the cross-linked thiol-ene polymer from the substrate. After removing the PS spheres, the metallic nanobowl structure with the thiol-ene substrate was achieved. In this paper, we fabricated gold nanobowl arrays composed of nanoheaves with the average diameter of ca. 60 nm between nanobowls, which verified the feasibility of the fabrication method.

INTRODUCTION

Nanostructures have been paid increasingly more attention due to exhibiting great potential applications in the areas of photo-catalysis[1], biosensors [2], magnetic [3] [4] [5] [6] [7] [8], and super-hydrophobic materials [9]. Advanced lithographic techniques currently being used to fabricate nanostructures include photolithography, electron beam direct writing, focused ion beam technology, and proximal-probe lithography[10] [11] [12] [13] [14] [15]. These techniques are capable of fabricating high-resolution features, but low cost and high volume for practical commerce still leads to great challenges.

The conglutination process is used widely in mechanical conglutination to stick one element to another element using adhesive material in industry[16]. Hence, adhesive material is the crucial component for conglutination process. Some research groups reported nanostructures could be fabricated via conglutination process based on different adhesive material. Truong*et al.* [17] described composite patterning elements that use a-PFPE as the features layer on the poly(dimethylsiloxane) (PDMS) or polyethylene terephthalate

(PET) support. The PDMS or PET film was used as the adhesion material to stick the a-PEPF nanostructures peeling off from the substrate.Chen*et al.* [18] reported a 2-D metallic nanobowl array on a thermoplastic substrate was fabricated using a piece of thermoplastic polycarbonate (PC) as adhesive material to be welded to the CNT film with metallic nanobowl array by microwave heating. Xu*et al.* [19] demonstrated the fabrication of nanobowl arrays composed of silver nanoparticle with an average diameter of ca. 10 nm via sphere lithography. As the adhesive material, a tape was stuck on the top of polystyrene (PS) spheres coated with silver film and then peeled off from the substrate. Therefore, the conglutination fabrication method was very simple, large-area, and low-cost, which provided a good idea for fabricating nanostructures in an economic and high-fidelity fashion. However, the common adhesive materials suffered some disadvantages, such as low adhesion, soft, and non-high-temperature resistance, to lead to the deformation or even damage of nanostructures and its distributions, which included glue, tape, and thermosetting materials such as PDMS, PET, and PC.

In 2001, Sharpless *et al.* [20] described a new concept for conducting organic reactions called "click reactions" to highly selectively form carbon heteroatom-linked molecular systems (C-X-C) with high efficiency under a variety of undemanding conditions. Hoyle and Bowman[21] reviewed the radical-mediated thiol–ene reaction as one click reaction. Because of the excellent mechanical and physical properties, thiol–ene polymer has been used widely as the nanoimprint resist in nanoimprint lithography (NIL) to fabricate high resolution nanostructures[22] [23] [24].

Herein, we proposed a simple and low-cost method to fabricate metallic nanobowl array via conglutination process based on the

UV-curable thiol–ene. The low viscous thiol–ene prepolymer was firstly coated on the self-assembly PS spheres template with metallic layer by confecting the thiol–ene monomers. When exposed under UV light, the prepolymer formed a cross-linked polymer network with high Young's modulus and surface energy to stick firmly to the metallic layer and peel off the PS sphere template from the substrate. The metallic nanobowl structure was obtained after dissolving the PS spheres into corrosive solution. Based on the new method, a gold nanobowl array with nanoheaves with the average width of ca. 60 nm between nanobowls was fabricated. The experimental results demonstrated that the fabrication method was feasible.

EXPERIMENTAL DETAILS

The experimental approach is schematically presented in Fig. 1. A monolayer of nanospheres was firstly self-assembled on a glass substrate inFig. 1(a). A metallic layer was then deposited on the nanospheres template inFig. 1(b). A UV-curing prepolymer with high adhesion and Young's modulus was spin-coating onto the metallic layer to form a thick film inFig. 1(c). When exposed under UV radiation in benign ambient, the prepolymer formed a cross-linked polymer network and stuck firmly to the metallic layer because of the high adhesion inFig. 1(d). The cross-linked polymer was then peeling off from the glass substrate with the metallic layer and PS spheres template inFig. 1(e). The PS spheres could be removed in the corrosion resolution and the metallic nanobowl structures on the polymer substrate were achieved inFig. 1(f). Because the cross-linked polymer was very rigid, the nanobowl structure was keeping the original arrangement of PS spheres.

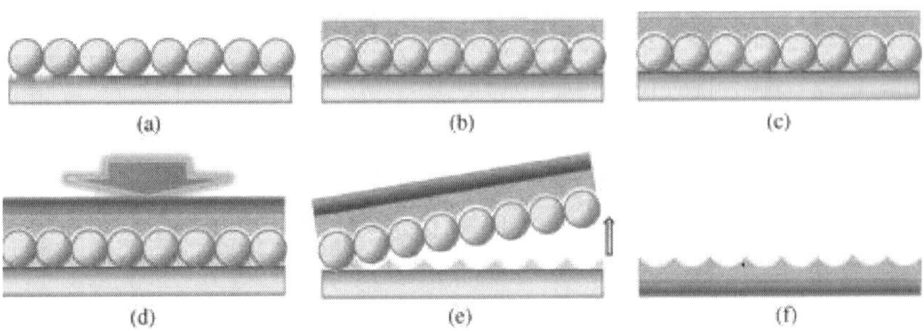

Fig. 1. Fabrication process of the metallic nanobowl array. (a) Self-assembly layer of nanospheres on glass wafer. (b) Depositing a metallic layer onto the nanospheres template. (c) Spin-coating UV-curing prepolymer film onto the metallic layer. (d) Curing by exposing under UV irradiation and form the cross-linked polymer network. (e) Peeling off the nanospheres template adhering to the metallic layer and polymer film. (f) Removing the nanospheres and achieve the metallic nanobowl array on the polymer substrate.

2.1. Materials and Instruments

Monodispersed 10 wt% polystyrene (PS) spheres with a diameter of 520 nm and 1-dodecanethiol were obtained from Duke. Tetrahydrofuran (THF) solution was purchased from commercial sources. Deionized water was used directly from Aquapro Company. Thiol–ene monomer materials such as poly(mercaptopropylmethyl siloxane (PMMS, 2,2-Dimethory-2-phenylacetopheneone (DMPA, and ethylene glycol dimethacrylate (EGDMA were purchased from MERYER. The chemical structures of the thiol–ene monomers are shown inFig. 2. The thiol–ene polymer was curing under UV irradiation and used to be adhesive material because of modulated physical and mechanical properties.

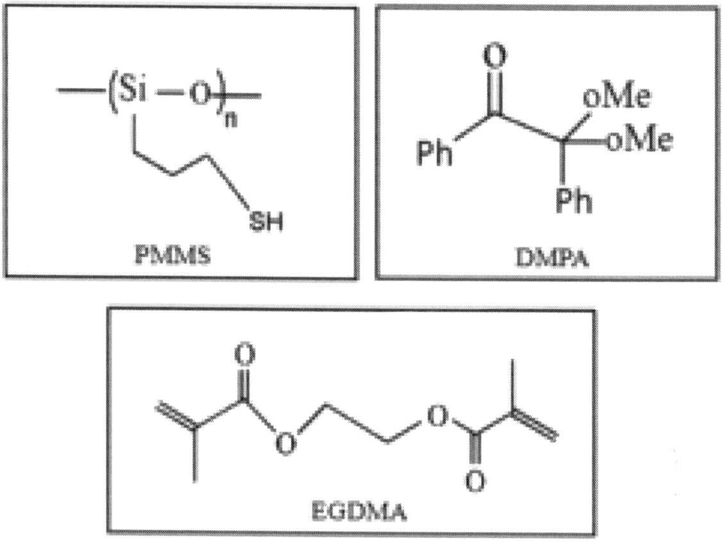

Fig. 2. Chemical structures of PMMS, DMPA, and EGDMA.

The glass wafers of mm were firstly hydrophilic-treated in mixture solution at 80 in a water bath for about one hour and cleaned by the deionized water. The glass wafers was then rinsed in solution for about one hour. Finally, they were cleaned with deionized water and dried with nitrogen. The cupping machine and dynamic contact angle instrument were used to measure the Young's modulus and surface energy, respectively.

2.2. Self-Assembly of PS Spheres

The PS spheres monolayer was prepared on the cleaned glass substrate via a self-assembling process [25]. The 10 wt% PS spheres solution was first diluted with an equal volume of deionized water and ultrasonicated for 10 s to improve the mixing. A 10- PS spheres mixed suspension was then assembled on the glass substrate by spin-coating at 2000 rpm for 25 s. The highly ordered template of the PS spheres was fabricated with the glass substrate after being dried at room temperature.

2.3. Deposition of Gold Film

A 100 nm thick gold film was deposited on the PS spheres template using a thermal evaporation system at a pressure of Pa. The spheres were embedded firmly into the gold film.

2.4. Fabrication of UV-Curing Polymer Substrate

The UV-curing thiol–ene prepolymer was coated onto the self-assembled nanospheres template, which was the compositions of PMMS (41.5 wt%), EGDMA (58.1 wt%), and DMPA (0.4 wt%) as the photoinitiator. Because of the low viscosity, the thiol–ene was dripped onto the nanospheres template to form a uniform film with thickness of 1 mm.Upon curing completely under a UV-LED lamp with the wavelength of 365 nm at 40 for 1 min, the thiol–ene prepolymer formed a cross-linked polymer network with the properties of high Young's modulus and high adhesion to the metal film[26], [27]. The Young's modulus of the cross-linked polymer was 5.5 GPa through measured by the cupping machine. The measured water contact angle of the cross-linked thiol–ene was 40° using dynamic contact angle instrument, showing that the surface energy was high to adhere firmly to the gold layer. The cross-linked thiol–ene polymer filled into the spaces among nanospheres and stuck firmly to the gold film. Because the adhesion between cross-linked polymer film and gold film was stronger than that between nanospheres template and glass substrate, the PS nanospheres template was better peeling off from the glass substrate with gold film and polymer substrate.

2.5. Removing of PS Spheres

To achieve the metal nanobowl structure, the PS spheres were removed with THF resolution. The PS spheres embedded into the gold film with polymer substrate was firstly immersed in a 5 mL THF solution for 2 min and then thoroughly rinsed into the ultrapure water. The PS spheres were dissolved in the THF solution and the gold nanobowl array with polymer substrate was obtained.

2.6. Characterization

Scanning electron microscopy (SEM) images were measured using a Hitachi S-4800 field-emission SEM at the primary electron energy of 3 kV. In order to observation to improve conductivity, the samples were sputtered with a thin gold layer (ca. 10 nm).

RESULTS AND DISCUSSIONS

The SEM images show the experimental results of each step described above in Fig. 3. As shown in Fig. 3(a), the monolayer of self-assembled PS spheres was coated with the gold film in a hexagonal close-pace structure, which was compact and uniform. The space between the spheres was partially filled as a result of the gold evaporation. Subsequently, the PS spheres coated with gold were peeled off from the glass substrate by applying a cross-linked thiol–ene polymer onto the top of the gold film. A top view of the transferred 520 nm diameter PS spheres with gold film and polymer substrate is shown inFig. 3(b). A large-area gold nanobowl array was created after removing the PS spheres inFig. 3(c). The nanoscale gaps among the nanospheres were filled with the gold to become the nanoheaves among the nanobowls. Thus, the nanobowl array was composed of gold nanoheaves with an average size of ca. 60 nm, as depicted inFig. 3(d). Due to the high rigid cross-linked polymer network, the nanobowl positions were unchanged and the coupling nanoheaves were uniformly distributed. The fabrication results indicated that the nanobowl array was high-precision and high-resolution, which demonstrated the feasibility of the novel fabrication method.

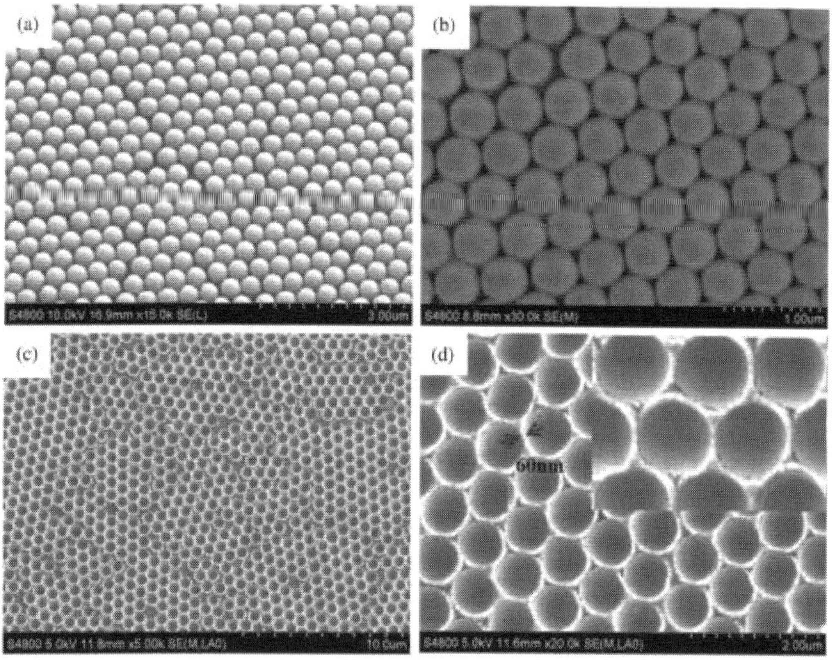

Fig. 3. SEM images of experimental results. (a) The monolayer of PS spheres coated with a gold film, (b) the transferred PS spheres with gold film and polymer substrate, (c) the generated gold nanobowl array on the cross-linked polymer substrate, and (d) an enlarged SEM image of the gold nanobowl array.

Some triangle spaces between the neighboring three nanobowls were not filled with metal to lead to some nanobowls distributed separately, because the PS nanospheres were not completely distributed compactly during the self-assembly process. Due to the thin metal film, some large gaps weren't filled with metal to lead to the separate nanobowls, which could be improved through optimizing the PS spheres self-assembly process.

The sizes of the nanobowl and nanoheaves between nanobowls can be modulated independently by selecting PS sphere with different diameters or by decreasing the PS sphere size with oxygen plasma etching after the self-assembly process. Therefore, the procedure of nanobowl fabrication could be extended to a wide range of applications with different self-assembled structures, sizes, and coating materials.

CONCLUSION

In summary, we developed a nonconventional method to fabricate metallic nanobowl array based on thiol–ene polymer via conglutination process, which relied on the high adhesive material to "stick" nanostructures. The low viscous thiol–ene prepolymer was coated on the self-assembly PS spheres template with metallic film. When exposed under UV light, the prepolymer formed a cross-linked polymer network with high Young's modulus and surface energy to stick highly to the metallic film. After peeling off from the glass substrate and dissolving the PS spheres in the corrosive solution, the metallic nanobowl array was fabricated. Using the novel method, a gold nanobowl array with ca. 60 nm nanoheaves was fabricated, which has great potential in applications of photonics, photo-catalysis, and magnetics.

REFERENCES

1. M. C. Wanke, O. Lehmann, K. Muller, Q. Wen and M. Stuke "Laser rapid prototyping of photonic band-gap microstructures" *Science, vol. 275, no. 5304, pp. 1284-1286, Dec., 1997*

2. A. J. Haes and R. P. V. Duyne "A nanoscale optical biosensor: Sensitivity and selectivity of an approach based on the localized surface plasmon resonance spectroscopy of triangular silver nanoparticles" *J. Amer. Chem. Soc., vol. 124, no. 35, pp. 10 596-10 604, Aug., 2002*

3. Y. N. Kim "Large magneto resistance in three dimensionally ordered macroporous perovskite manganites prepared by a colloidal templating method" *J. Mater. Chem., vol. 14, no. 11, pp. 1774-1777, May, 2004*

4. L. Xu "Synthesis and magnetic behavior of nickel sphere arrays" *Adv. Mater., vol. 15, no. 18, pp. 1562-1564, Sep., 2003*

5. E. O. Chi, Y. N. Kim, J. C. Kim and N. H. Hur "A macroporous perovskite manganite from colloidal templates with a curie temperature of 320 K" *Chem. Mater., vol. 15, no. 10, pp 1929-1931, Apr., 2003*

6. L. Xu "Electrodeposited nickel and gold nanoscale metal meshes with potentially interesting photonic properties" *Chem Commun., no. 12, pp. 997-998, May, 2000*

7. J. C. Kim, Y. N. Kim, N. H. Hur, W. S. Kim and Y. G. Kang "Highly ordered macroporous magnetic materials prepared by electrodeposition through colloidal template" *Phys. Status Solidi., vol. 241, no. 7, pp. 1585-1588, May, 2004*

8. J. W. Lau and J. M. Shaw "Magnetic nanostructures for advanced technologies: Fabrication, metrology and challenges" *J. Phys. D: Appl. Phys., vol. 44, no. 30, pp. 303001-1-303001-43, Jul., 2011*

9. W. G. Xu "Preparation of superhydrophobic ZnO films on zinc substrate by chemical solution method" *Chem. Res. Chin. Univ., vol. 28, no. 3, pp. 529-533, Aug., 2012*

10. R. F. Pease and S. Y. Chou "Lithography and other patterning techniques for future electronics" *Proc. IEEE, vol. 96, no. 2, pp. 248-270, Feb., 2008*

11. D. Xia, Z. Ku, S. C. Lee and S. R. Brueck "Nanostructures and functional materials fabricated by interferometric lithography" *Adv. Mater., vol. 23, no. 2, pp. 147-179, Jan., 2011*

12. E. Reznikova, J. Mohr, M. Boerner, V. Nazmov and P.-J. Jakobs "Soft X-ray lithography of high aspect ratio SU8 submicron structures" *Microsyst. Technol., vol. 14, no. 9–11, pp. 1683-1688, Oct., 2008*

13. M. Saito and J. Taniguchi "Electron beam direct writing of nanodot patterns on roll mold surfaces by electron beam

on-off chopping control" *Microelectron. Eng., vol. 123, pp. 89-93, Jul., 2014*

14. A. Łaszcz, W. Nogala, A. Czerwinski, J. Ratajczak and J. Kątcki "Fabrication of electrochemical nanoelectrode for sensor application using focused ion beam technology" *Polish J. Chem. Tech., vol. 16, no. 3, pp. 40-44, Nov., 2014*

15. N. V. Doroshovets "0.1-nanometer resolution positioning stage for sub-10 nm scanning probe lithography" *Proc. SPIE/COS, vol. 8680, 2013*

16. S. Hao, Y. Tan, L. Gao, X. Wang and Z. Qian "Application of conglutination technology in military equipment" *Mech. Eng. Autom., vol. 4, no. 2, pp. 214-216, Apr., 2012*

17. T. T. Truong "Soft lithography using acryloxy perfluoropolyether composite stamps" *Langmuir, vol. 23, no. 5, pp. 2898-2905, Jan., 2007*

18. T. H. Chen "Two-dimensional metallic nanobowl array transferred onto thermoplastic substrates by microwave heating of carbon nanotubes" *Nanotechnol., vol. 19, no. 46, Oct., 2008*

19. M. Xu "Fabrication of functional silver nanobowl arrays via sphere lithography" *Langmuir, vol. 25, no. 19, pp. 11 216-11 220, Sep., 2009*

20. H. C. Kolb, M. G. Finn and K. B. Sharpless "Click chemistry: Diverse chemical function from a few good reactions" *Angew. Chem. Int. Ed., vol. 40, no. 11, pp. 2004-2021, May, 2001*

21. C. E. Hoyle and C. N. Bowman "Thiol–ene click chemistry" *Angew. Chem. Int. Ed., vol. 49, no. 9, pp. 1540-1573, Feb., 2010*

22. H. Lin, X. Wan, X. S. Jiang, Q. K. Wang and J. Yin "A nanoimprint lithography hybrid photoresist based on the thiol–ene system" *Adv. Funct. Mater., vol. 21, no. 15, pp. 2960-2967, Aug., 2011*

23. H. Lin, X. Wan, X. Jiang, Q. Wang and J. Yin "A "thiol–ene" photo-curable hybrid fluorinated resist for the high-performance replica mold of Nanoimprint Lithography (NIL)" *J. Mater. Chem., vol. 22, no. 6, pp. 2616-2623, Dec., 2012*

24. M. Zhang "Fabrication of nano-pillar with sub-100 nm resolution based on thiol–ene"

25. *Proc. SPIE/COS, vol. 9277, 2014, Beijing, China*

26. Y. S. Jung, J. B. Chang, E. Verploegen, K. K. Berggren and C. A. Ross "A path to ultranarrow patterns using self-assembled lithography" *Nano Lett., vol. 10, no. 3, pp. 1000-1005, Feb., 2010*

27. P. Jin, N. Liu, J. Lin, J. B. Tan and P. D. Prewett "Replication of micro-optical elements with continuous relief by ultraviolet embossing with thiol–ene-based resist" *Appl. Opt., vol. 50, no. 21, pp. 4063-4067, Jul., 2011*

28. H. Wook Ro "Cubic silsesquioxanes as a green, high-performance mold material for nanoimprint lithography" *Adv. Mater., vol. 23, no. 3, pp. 414-420, Aug., 2011*

5

CHAPTER

Post-Plasma SiO_x Coatings of Metal and Metal Oxide Nanoparticles for Enhanced Thermal Stability and Tunable Photoactivity Applications

Patrick Post [1], Nicolas Jidenko [2], Alfred P. Weber [1,*] and Jean-Pascal Borra [2]

[1]Institute of Particle Technology, Clausthal University of Technology, Leibnizstrasse 19, 38678 Clausthal-Zellerfeld, Germany

[2]Lab of Phys Gaz and Plasmas, CNRS, Univ. Paris Sud, CentraleSupelec, Université Paris-Saclay, F-91405 Orsay, France

ABSTRACT

The plasma-based aerosol process developed for the direct coating of particles in gases with silicon oxide in a continuous chemical vapor deposition (CVD) process is presented. It is shown that non-thermal plasma filaments induced in a dielectric barrier discharge (DBD) at atmospheric pressure trigger post-DBD gas phase reactions. DBD operating conditions are first scanned to produce ozone and dinitrogen pentoxide. In the selected conditions, these plasma species react with gaseous tetraethyl orthosilicate (TEOS) precursor downstream of the DBD. The gaseous intermediates then condense on the surface of nanoparticles and self-reactions lead to homogeneous solid SiO_x coatings, with thickness from nanometer to micrometer. This confirms the interest of post-DBD injection of the organo-silicon precursor to achieve stable production of actives species with subsequent controlled thickness of SiO_x coatings. SiO_x coatings of spherical and agglomerated metal and metal oxide nanoparticles (Pt, CuO, TiO_2) are achieved. In the selected DBD operating conditions, the thickness of homogeneous nanometer sized coatings of spherical nanoparticles depends on the reaction duration and on the precursor concentration. For agglomerates, operating conditions can be tuned to cover preferentially the interparticle contact zones between primary particles, shifting the sintering of platinum agglomerates to much higher temperatures than the usual sintering temperature. Potential applications for enhanced thermal stability and tunable photoactivity of coated agglomerates are presented.

Keywords: particle coating; silicon oxide; dielectric barrier discharge; gas phase; continuous process; TEOS; CVD

1. INTRODUCTION

The coating of nanoparticles is an essential step to improve the particle thermostability [1], to create a protective layer between particle and environment [2] or to enhance particle dispersion in liquids. Silicon oxide is one of the preferred nontoxic and almost inert coating materials.

Many coating techniques use liquid phase reactions with additional steps such as washing, drying and separation [3]. Moreover, impurities from solvents remain in the final solid coatings. Therefore, gas phase coating techniques are generally preferred [4] (e.g., for atomic layer deposition [5], for flames [6,7] and for plasma enhanced chemical vapor deposition (CVD) [8]).

Most publications on SiO_x coatings deal with the coating of flat surfaces, but few groups used it to coat nanoparticles directly in the gas phase. Homogeneous coatings can be performed in non-continuous devices such as fluidized bed reactors [9]. In flame synthesis, direct coating of gas born nanoparticles such as TiO_2, may be performed in a continuous one-step process.

A more versatile coating method is the CVD that can be induced by photochemistry [10] and plasmas. In dielectric barrier discharges (DBD), plasma filaments produce reactive species (electrons, photons, radicals, excited species and more stable ozone and nitrogen oxides in air), triggering the polymerization of organic and organo-metal precursors. For the coating of substrates, the precursors can be injected in the DBD or in post-DBD for SiO_x coatings [8,11], as well as for functional polymer coatings [12,13]. For the coating of supported particles by plasma, Mori *et al.* [14], Kogoma *et al.* [15] and Brüser *et al.* [16] used batch processes.

The direct SiO_x coating of suspended nanoparticles in gases using atmospheric pressure DBD is an emerging field [13]. Nessim *et al.* used HMDSO (hexamethyldisiloxane) and TEOS for the coating of particles either directly in a plasma as Vons *et al.* [17], or in post-DBD [18]. As for flat surface coating, the precursor injection in the DBD leads to a loss of functionality of the gaseous precursor reacting with plasma species in the gap, to particles electro-collection on surfaces in the gap and to discharge destabilization that hampers the economics of the process.

Here, post-DBD injection of the organo-silicon precursor is tested to avoid electrode coating so as to achieve stable DBD production of active species with subsequent controlled thickness of SiO_x coatings of nanoparticles *versus* tetraethyl orthosilicate (TEOS) concentration and reaction duration. The setup and operating conditions are presented. Preliminary tests confirm that plasma species trigger the post-DBD conversion of TEOS into solid SiO_x coatings of nanoparticles. DBD operating conditions are then scanned to produce ozone and dinitrogen pentoxide. In the selected conditions, subsequent sections present the coating of spherical particles and of agglomerates separately *versus* reaction duration and concentrations of reactants (plasma species and TEOS). Finally, properties of coated agglomerates for thermal stabilization of catalyst particles and for photoactivity control are depicted.

2. EXPERIMENTAL SECTION

The experimental setup is shown in Figure 1. The first step is the production of nanoparticles with defined size, concentration and morphology (spheres or agglomerates). Then, a controlled amount of precursor is added. The third step is the post-DBD injection of this TEOS/nanoparticles aerosol, mixed with plasma species. Finally, post-DBD reaction times are tuned in different volumes.

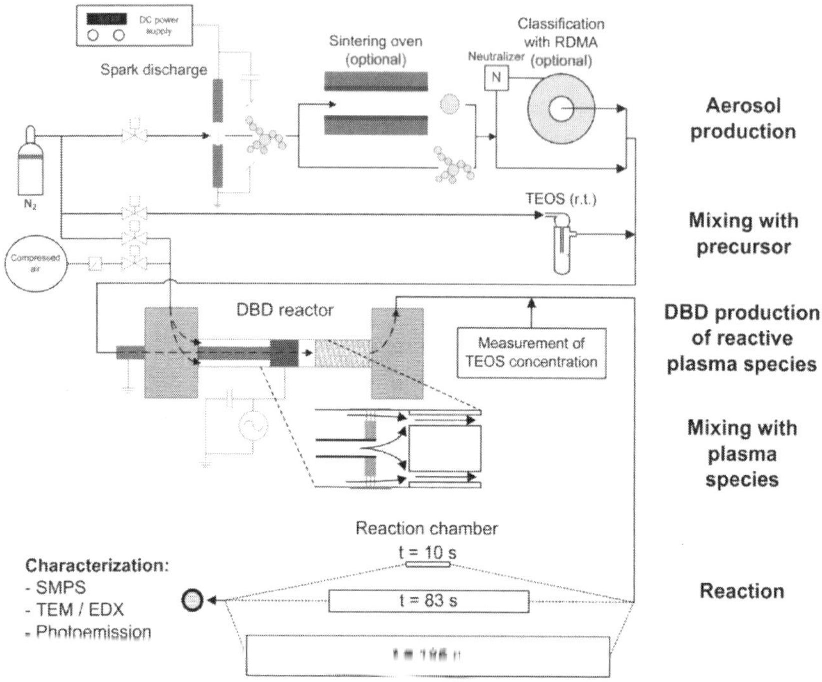

Figure 1. Experimental setup with the different process steps.

2.1. Aerosol Production and Mixing with Precursor

The metal core particles were produced in a spark discharge generator (SDG) without any side products in the gas stream [19]. Two electrodes of the same material with a 5 mm gap were polarized with a high voltage DC power supply. Then, repetitive sparks develop between the electrodes leading to vaporization and nucleation into nanoparticles [19]. The preferred material was platinum for its high TEM contrast and applications, e.g., in catalysis. Due to the high number concentrations of small primary particles large fractal agglomerates formed rapidly behind the SDG in 1 L/min N_2 (standard liter per minute, nitrogen 5.0). Figure 2 (left) shows such agglomerates, converted into spherical particles (right) by sintering in a tube furnace at 1000 °C. Pt, Cu, Ti, Ni and Fe have been used in the spark generator to produce similar nanoparticles, agglomerated or spherical after sintering, which could be size selected with a Radial Differential Mobility Analyzer (RDMA) downstream of an X-ray source to establish a known charge distribution.

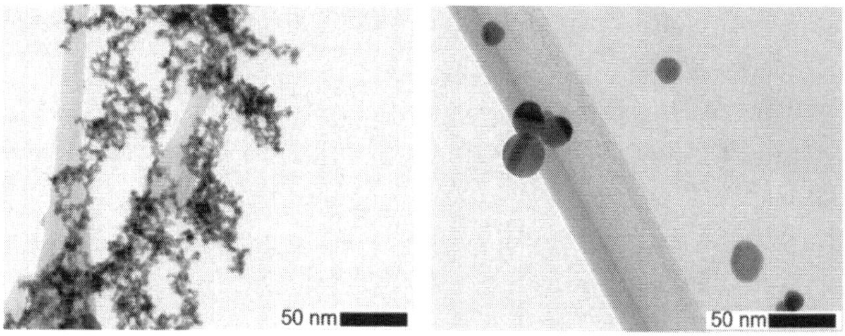

Figure 2. Effect of sintering on the morphology of platinum nanoparticles from the spark generator: (**left**) agglomerates downstream of the spark before sintering; and (**right**) spherical particles after sintering.

Among the organo-silicon precursors, tetraethyl orthosilicate (TEOS, $Si(OC_2H_5)_4$) is one of the most prominent. Due to its non-hazardous character, TEOS is easy to handle. Moreover, its relatively low vapor pressure facilitates the addition of small amounts of gaseous precursor and it reacts more slowly compared to other precursors and is therefore suitable to achieve nano-sized SiO_x coatings.

The mixing of the TEOS with the particles is done in a simple T-piece prior to post-DBD injection. A second flow of nitrogen below 30 mL/min is used to transport the precursor vapor from a bottle. The nitrogen does not bubble through the precursor but flows over the surface of the liquid to prevent the formation and transport of droplets. In the bottle kept at room temperature (24 °C), the distance between the gas inlet and the surface of the initially liquid TEOS is kept constant at 5 mm.

The precursor concentration was measured by Fourier transform infrared spectroscopy (FTIR, Tensor 27, Bruker Optik GmbH, Ettlingen, Germany) on the Si–O absorbance near 1100 cm^{-1}, directly after the mixing of all gas flows downstream of the DBD reactor (Figure 1), *i.e.*, at the entrance of the post-DBD reaction chamber. It can be tuned from 0.3 to 4.1 ppmv (below 0.2% of TEOS saturation vapor pressure of 242 Pa at 24 °C), without affecting significantly the total flow of 3 L/min (plus a maximum of 30 mL/min of the N_2-TEOS mixture).

2.2. DBD Arrangement and Operating Conditions

The DBD is made of two stainless steel parts that hold a quartz glass tube with an inner diameter of 13 mm and a thickness of 1.5 mm serving as dielectric (see inset of Figure 1). On the outside of the glass tube, an electrode made of copper adhesive tape is connected to the high voltage AC power supply. A stainless steel disc (diameter: 12 mm, length: 2 mm) is fixed on the grounded central stainless steel tube in front of the polarized external electrode. Plasma filaments then occur in the 0.5 mm ring-shaped gap between the metal disc and the dielectric, in a so-called monoDBD [20].

To limit the electrode temperature below 100 °C, the generator frequency was fixed to 41 kHz. The voltage was measured with a HV probe. The current pulses related to each plasma filament were recorded via a 50 Ω input resistor to evaluate the charge and the energy per filament as well as the power, calculated, with n the number of periods and T the period duration, according to:

$$P = \frac{1}{n \cdot T} \cdot \int_{nT} u(t) \cdot i(t)\, dt \qquad (1)$$

2.3. Mixing with Plasma Species and Reaction

The particles and the precursor flow through a stainless steel tube (4.57 mm inner diameter) without contact with the plasma. The combined aerosol flow from the spark generator and the TEOS bottle was injected into the inner tube at 1 L/min N_2 varying only slightly with the small amount of precursor gas flow. A 1:1 mixture of nitrogen and filtered air was injected at 2 L/min into the DBD gap to transport plasma species downstream of the DBD up to the post-DBD mixing with the aerosol/precursor mixture. The total flow of about 3 L/min was injected in the post-DBD reaction volume. To enhance the mixing of plasma species and of the particle/TEOS aerosol flow, a static mixer was placed about 8 mm downstream of the DBD so that there is no reflux of TEOS into the discharge zone. As shown in Figure 1, it deflects the inner flow towards smaller tubes near the outer diameter of the mixer, leading to turbulent mixing of the two gas flows.

The influence of the reaction duration was tested with different tubes as reaction chambers with laminar flow leading to defined transit time distributions. Unless stated otherwise, the selected conditions defined from the preliminary results detailed in Section 3.2, are specified in Table 1.

Table 1. Summary of the standard experimental conditions.

Description	Abbr.	Value
Volume flow through discharge	Q_{Plasma}	2 L/min
Volume flow in particle generation step	$Q_{Particles}$	1 L/min
TEOS concentration	c_{TEOS}	0.7 ppmv
Oxygen concentration in discharge	c_{Oxygen}	10.5%
Applied peak-to-peak voltage	U_{PP}	10 kV
Applied plasma power (Figure 3)	P_{Plasma}	4 W
Reaction time before sampling	$t_{Reaction}$	83 s

2.4. Aerosol Characterization

Downstream of the reaction chamber, the coated particles were collected at 0.2 L/min in a bypass across a lacey transmission electron microscopy (TEM) grid for 2 min. As a consequence, the particles analyzed with the TEM refer to different ages since condensation and heterogeneous reactions still happen during the 2 min collection. However, for the presentation of the experimental data, the beginning of the sampling was chosen as time reference. Then, the TEM grid was removed and stored in ambient air for at least 20 min before being transferred under ultra-high vacuum into the TEM. Coupled with the energy-dispersive X-ray spectroscopy (EDX) sensor, size, coating thickness and atomic composition of the coated particles were characterized.

On-line measurements of the particle number size distributions were obtained with a scanning mobility particles sizer (SMPS Grimm 5.403, GRIMM Aerosol Technik GmbH & Co. KG, Ainring, Germany). Assuming spherical particles, the coating thickness is defined as half the difference of the mobility equivalent mode diameters of non-coated and coated particles.

The coated surface of particles was estimated by aerosol photoemission measurements (APE) [21]. Downstream of an electrostatic precipitator, the neutral aerosol exposed to UV is charged positively due to electron emission from metal nanoparticles. The aerosol charge, measured with a Faraday Cup Electrometer after an ion trap is proportional to the uncoated active surface area.

3. RESULTS AND DISCUSSION

3.1. DBD Electrical Characterisation and Post-DBD Ozone and NO_x

As shown in Figure 3 (left), both voltage and current are sinusoidal with a phase shift between them. The real discharge current with pulses related to each plasma filament is superposed to this sinusoidal capacitive current. From the instantaneous voltage and current, the discharge power was calculated as detailed above and depicted in Figure 3 (right) *vs.* applied voltage.

At increasing voltages, the plasma filaments originate from the edges of the inner electrode until they cover the whole surface of the smaller electrode for higher applied voltages (see insets of Figure 3). As expected for such asymmetrical mono-DBD [20], negative current pulses for filaments developing from the metal electrode to the dielectric surface are slightly larger than positive pulses related to the same streamer-like plasma filament, then directed towards the central metal electrode.

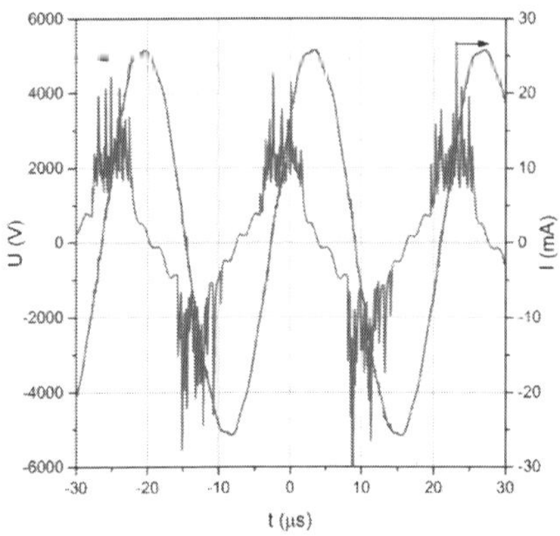

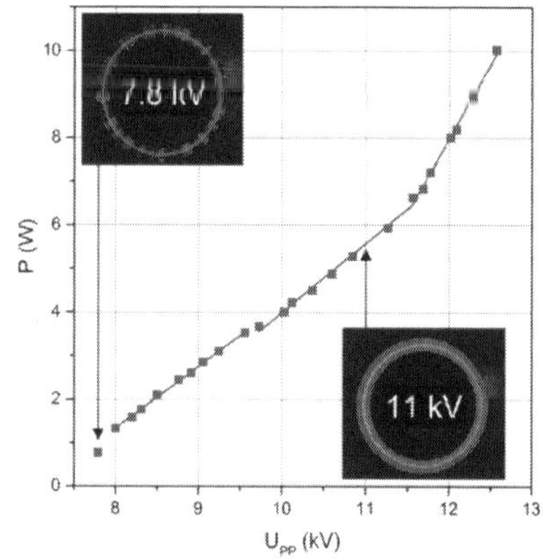

Figure 3. Temporal evolution of voltage and dielectric barrier discharge (DBD) current (**left**) and DBD power *versus* peak-to-peak voltage characteristic of the DBD with insets of violet light emitted from plasmas in air (**right**).

Additional post-DBD FTIR analysis of gaseous products without any TEOS reveal ozone and NO_x, respectively, from 0 to 301 ppmv for ozone and from 35 to 114 ppmv for NO_2, N_2O, N_2O_5 and HNO_x, without NO, as depicted in Table 2 for different voltages. The species were quantified from the adsorption signal at the corresponding wavelengths of each species: O_3: 1050 cm^{-1}, NO_2: 1630 cm^{-1}, N_2O: 2340 cm^{-1}, N_2O_5: 1240 cm^{-1}, NO: 1890 cm^{-1} and HNO_x: 1330 cm^{-1}. These species were expected from a DBD fed with a 10% O_2/90% N_2 mixture, since these DBD were used as ozonizers.

Table 2. Electrode and reaction temperatures, and ozone and NO_xconcentrations (in ppmv) for three DBD voltages. * HNO_x could not be quantified due to significant interference.

$u_{DBD, PP}$ (kV)	$T_{electrode}/T_{reaction}$ (°C)	O_3	NO_2	N_2O	N_2O_5	NO	HNO_x	Total NO_x
8	33/20	267 ± 9	23 ± 8	2 ± 2	10 ± 0	0	*	35
10	46/35	301 ± 3	20 ± 0	20 ± 4	24 ± 0	0	*	64
12	120/<70	0	76 ± 5	38 ± 4	0	0	*	114

Finally, it can be noted from electrode temperatures depicted in Table 2 that up to 10 kV the gas is not heated above 100 °C in the gap of the DBD since ozone destruction would be faster than its production. On the contrary, at 12 kV, ozone and dinitrogen pentoxides are converted into other NO_x, as expected for higher gas temperatures arising from higher electrode temperatures.

In the selected arrangement, a stable DBD with constant energy and number of pulses per cycle leads to constant related post-DBD concentrations of plasma species at the entrance of the reaction volume. These concentrations can be tuned with DBD operating conditions with ozone and N_2O_5 up to 10 kV or only with other nitrogen oxides at 12 kV, for reaction temperatures near ambient temperatures up to 35 °C for 8 and 10 kV and below 70 °C at 12 kV, respectively.

3.2. Proof of Concept and Ranges of TEOS Concentration and Reaction Duration for Nanoparticle Coating

It is first confirmed that post-DBD gas phase reactions of ozone and dinitrogen pentoxide plasma products with gaseous TEOS precursor lead to final homogeneous SiO_x coatings.

This can be stated from EDX analysis presented below in Table 3 and from Figure 4 (left). The size distributions of sintered particles do not evolve by mixing with TEOS precursor unless plasma filaments develop in the DBD. Then, a significant increase in diameter is found when the DBD is fed with a 10% O_2/90% N_2 mixture (Figure 4, left), while in pure nitrogen, *i.e.*, without oxygen injected either in the DBD or in post-DBD, some holey heterogeneous coatings are formed. As sub-second transit time from the DBD gap (where plasma filaments produce electrons, photons, radicals, excited species) to the entrance of the reaction volume is longer than the times of life of these plasma species (e.g., for atomic oxygen—10^{-8} s for O^1d, 10^{-8} to 10^{-7} s for O^3p- and metastable $O_2{}^1\Delta g$ and vibrationally excited N_2, up to 10^{-3} s, in air [22]), only stable species such as ozone and nitrogen oxides are analyzed in post-DBD at the entrance of the reaction volume and could react with TEOS.

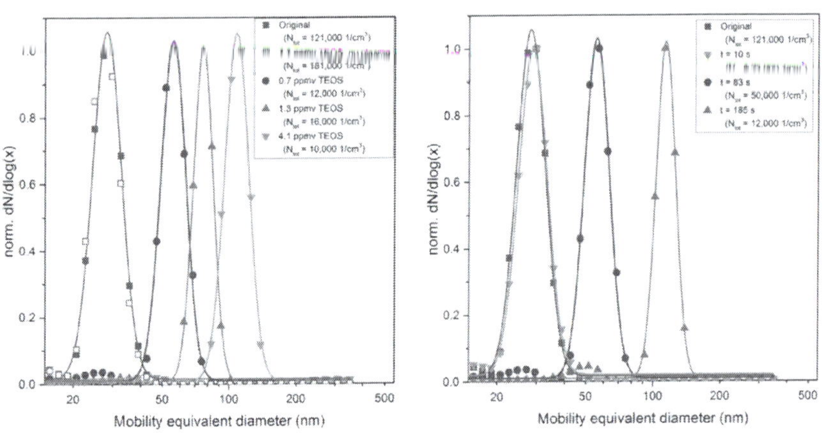

Figure 4. Mobility equivalent diameter measured with scanning mobility particles sizer (SMPS): (**left**) as a function of the tetraethyl orthosilicate (TEOS) concentration at constant reaction time $t_{Reaction}$= 83 s; and (**right**) as a function of reaction time at constant TEOS concentration of 0.7 ppmv (the total number concentration N_{tot} is given in the legend).

Table 3. Elemental composition from the quantitative energy-dispersive X-ray spectroscopy (EDX) analysis as a mean of several measurements on the samples for two different reaction times.

Element	$t_{Reaction}$ = 83 s	$t_{Reaction}$ = 185 s
Si	5 at.%	23 at.%
O	13 at.%	48 at.%
C	30 at.%	23 at.%
Pt (core)	53 at.%	6 at.%

Actually, Okuyama *et al.* [23] and Fujino *et al.* [24] studied the kinetics of TEOS and ozone reactions, which was later modeled by Romet *et al.* [25]. Atomic oxygen produced by ozone decomposition is suspected to be the main starter of the reaction [25]. One critical step is the gas phase conversion of TEOS into triethoxysilanol ($Si(OC_2H_5)_3OH$), then condensed on the surface and finally converted into solid SiO_x. A simple model may describe the overall reaction between TEOS and ozone as [25]:

$$Si(OC_2H_5)_4 + 4\,O_3 \rightarrow SiO_2 + 4\,CH_3CHO + 2\,H_2O + 4\,O_2$$

The coatings formed at different voltages are shown in Figure 5. Solid SiO_x are formed up to 10 kV, contrary to 12 kV when ozone and N_2O_5 are not present anymore downstream of the DBD (Table 2 andFigure 5, right). Similar homogeneous 2 nm coatings at 8 and 10 kV probably result from the similar ozone concentrations for both voltages (Figure 5, left and center).

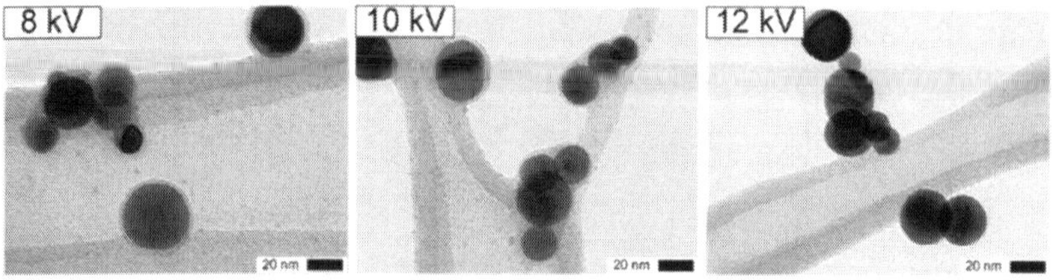

Figure 5. Transmission electron microscopy (TEM) micrograph of final solid coating of nanoparticles for different applied voltage (U_{pp}): **(left)** 8 kV; **(middle)** 10 kV; and **(right)**: 12 kV, for 0.4 ppmv TEOS and $t_{Reaction}$ = 185 s.

The importance of ozone is further corroborated by a decrease in its concentration when TEOS is added to the system. FTIR measurements analogous to those described above yield an ozone concentration of only 55 ppmv at 10 kV, which is significantly lower than without precursor (301 ppmv). On the other hand, the N_2O_5 concentration of 19 ppmv is comparable to that measured without TEOS (24 ppmv). The addition of Pt particles to the system shows no further change in the ozone concentration (55 ppmv), but N_2O_5 decreases somewhat to 15 ppmv. While these findings suggest that ozone is the most important species for the reaction with TEOS in this process, further research is required to understand all possible reaction pathways.

Finally, since atomic oxygen can still be produced from slow post-DBD reactions of ozone and N_2O_5, it is not yet clear if TEOS reacts directly with ozone or with atomic oxygen to form condensable intermediate species.

3.2.1. Coating Composition

For two coating thicknesses controlled by the reaction time (Table 3), the particles were analyzed by EDX showing contributions of Pt (core particles) and Cu from the TEM grid and sample holder as well as C, O and Si. While Si and O obviously belonged to the SiO_x coatings, the source of C at least partly arises from the carbon coating of the TEM grid. As expected, the ratio of the coating elements to the core material increases for thicker coatings. The ratio of Si to O is nearly stoichiometric SiO_2 with a slight excess of oxygen. To determine the actual carbon content, coated Pt particles were deposited on a Mo TEM grid without a C film. The experiment showed that there is indeed some amount of carbon in the coated particles. The chemical composition of the coating following from this measurement is $SiO_{2.3}C_{1.5}H_z$. The H content cannot be measured with EDX.

3.2.2. Ranges of TEOS Concentration and Reaction Duration for Nanoparticle Coating

With the selected DBD arrangement polarized with applied voltages up to 10 kV at 3 L/min, different TEOS concentrations and reaction times have been tested. The goal was to define the ranges of conditions for the parametric study to control the thickness of nanometer sized coating, detailed in the next sections.

For longer reaction durations, a 20 L tank was used with a broad residence time distribution, varying from a few minutes up to several tens of minutes. With this tank downstream of the DBD and the highest TEOS concentrations (1400 ppmv), high concentrations of micron-sized spherical dense SiO_xparticles were collected on TEM grids even without Pt nanoparticles from the spark generator (seeFigure 6, left.). In that case, post-DBD heterogeneous nucleation of condensable gaseous intermediates probably occurs on single digit nano-sized particles and/or on ions reported downstream of DBDs [22,26,27]. Lowering the TEOS concentrations and reaction time lead to much smaller sub-micron-sized SiO_x coating

(Figure 6, center), down to nanometer thick homogeneous coatings on 30 nm Pt nanoparticles (Figure 6, right).

Figure 6. TEM micrographs: (**left**) Large SiO_x particles (1.13 μm) formed in the gas phase during the reaction duration in the 20 L tank (U_{pp} = 10 kV, with post-DBD injection of c_{TEOS} = 1400 ppmv without Pt nanoparticles); (**middle**) smaller sub-micron sized SiO_xcoated Pt particles, formed and agglomerated in the gas phase (U_{pp}= 8.6 kV, 20 L tank with 280 ppmv TEOS and Pt nanoparticles), and (**right**) nanometer thick coatings of Pt nanoparticles (U_{pp} = 8 kV,$t_{Reaction}$ = 185 s with 0.4 ppmv TEOS).

For different reaction durations and TEOS concentrations, TEM and EDX analyses confirm that SiO_xparticles are always formed in the gas phase, even if slow reactions of SiO_x formation still occur after collection near ambient temperature, as proved in the next section. Such spherical SiO_x particles usually arise from liquid droplets [4,7].

As the precursor concentration is below saturation (<0.2% of $p_{sat\ TEOS}$), these droplets cannot be formed by TEOS condensation. Indeed, the size distributions measured on line downstream of the reactor with SMPS, are the same for spherical sintered Pt nanoparticle with or without TEOS (see Figure 4, left). These droplets are thus formed by condensation of gaseous intermediate species continuously formed by TEOS reactions in post-DBD, up to the onset of condensation.

Hence, nanoparticles can be coated with SiO_x by a three step process starting with gas phase reactions of TEOS with post-DBD ozone and dinitrogen pentoxides, producing intermediate species more condensable than the TEOS precursor. In a second step, droplets are formed by condensation of silanol-like intermediates on nanoparticles, then converted by sol-gel reactions into solid SiO_x coatings. Finally, coagulation discussed below sometimes leads to agglomerated coated particles.

As expected from kinetic considerations, the growth rate of coating and the related thickness depend on the reaction conditions, including temperature and reactants concentrations downstream of the DBD (ozone and N_2O_5 plasma species, as well as TEOS). From a practical point of view, for fixed DBD operating conditions, the reaction time and the precursor concentration control the SiO_x coating thickness. To avoid the coating of the TEM grid, low TEOS concentrations in the ppmv range will be used. In this way, nanometer coating thickness can be achieved, as detailed in the next sections.

3.3. Homogeneous Nanometer-Sized Coating of Spherical Nanoparticles

To investigate the influence of particle properties and operational parameters on the coating thickness, size-selected sintered spherical Pt nanoparticles were used. TEOS concentrations and the reaction time were varied and the coating thickness was determined for different materials.

As outlined above, during the reaction of TEOS vapor with the plasma species intermediate molecules are formed which condense on the particle surfaces. Since the Knudsen number

(Kn), defined as twice the mean free path ($\lambda \approx 65$ nm) of the gas molecules divided by the particle diameter, is significantly larger than 1 ($Kn \approx 4$), the condensation takes place close to the free molecular regime ($Kn >> 1$). In this case the theory for heterogeneous condensation [28] predicts that the particle diameter increases linearly with the concentration of the condensing vapor and with the residence time as long as the concentration of the condensing vapor is constant, *i.e.*, without depletion:

$$\Delta R = \frac{M \cdot (p - p_S)}{\rho_l \cdot N_A \cdot \sqrt{2\pi mkT}} \cdot \Delta t \qquad (2)$$

where ΔR = coating thickness, M = molecular weight of the liquid, p = partial pressure of condensing vapor, p_S = partial pressure at particle surface, ρ_l = liquid density, N_A = Avogadro number, m = mass of vapor molecule, and kT = thermal energy.

3.3.1. Kinetics of TEOS Conversion into SiO_x Coating

Although the mechanisms and the kinetics of the formation reactions leading to the condensing species are not known in detail, our results support that such depletion of intermediate condensable species does not affect condensation here, even though it necessarily happens along the reaction volume, while post-DBD plasma species are consumed by reaction with TEOS. However, without a final forced condensation step by cooling, all formed intermediates do not condense. The results shown inFigure 4 and in Figure 8 confirm that free molecular condensation is a reasonable approximation to describe the thickness of the liquid coating *versus* TEOS concentration and reaction time, as detailed below.

The size distribution of particles/drops suspended in carrier gas was first measured with the SMPS at ambient conditions before collection, *i.e.*, for the reference time of reaction, while TEM analyses were performed on particles with residual solid coating only, formed after additional time delay of 2 min maximum for particles collected immediately and at least 20 min in air before final forced evaporation under vacuum conditions for TEM analysis. Therefore, the TEM results show the layers having a vapor pressure low enough to withstand the vacuum, *i.e.*, solid coatings.

Some examples of TEM micrographs are shown in Figure 7 for a TEOS concentration of 0.7 ppmv and different reaction times. While the solid coating for a reaction time of 83 s is only 2.5 nm thick, it increases to 38 nm after 185 s (Figure 8, right). For longer reaction times, agglomeration may take place as shown in Figure 7 (right). Agglomeration may occur by diffusion or by electrical forces between oppositely charged particles, since bipolar ions are produced downstream of such a DBD, which can even be used as an aerosol neutralizer [26,27].

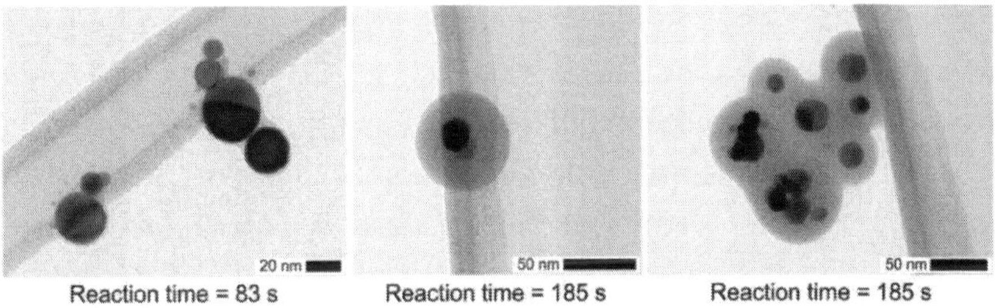

Figure 7. TEM micrographs showing the time dependence of the coating thickness for a constant TEOS concentration of 0.7 ppmv.

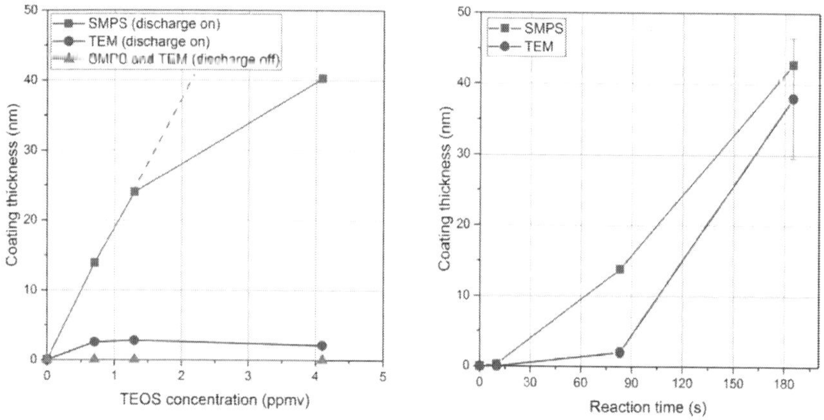

Figure 8. Coating thickness measured with scanning mobility particles sizer (SMPS) and TEM as a function of TEOS concentration for t_{Reaction} = 83 s (**left**) and as a function of the reaction time for 0.7 ppmv TEOS concentration (**right**).

Particles and a few particle-clusters are well separated with smeared boundaries so that the coating thickness may only be deduced from the outside contour. This supports that core particles are not mobile in the coating, *i.e.*, the core is surrounded by a solid or at least more viscous gel shell. However, at the moment of collision the surface is liquid-like leading to conforming contacts. This is confirmed from the larger coating of drops/particles (measured before collection with SMPS) in comparison with the residual solid coating, formed after additional time delay before the final forced evaporation under vacuum conditions for TEM analysis (Figure 8), detailed below.

For a reaction time of 83 s, the coating thickness measured with SMPS increases with the TEOS concentration in qualitative agreement with Equation (1) (Figure 8, left). The deviation from the linear relationship at higher TEOS concentration may indicate that the concentration of the condensing vapor species is not directly proportional to the TEOS concentration. The reaction time of 83 s seems too short to complete the conversion of all the liquid coating into a solid silica coating. The reaction of formation of solid SiO_x is thus not limited by the amount of intermediate (unless much smaller precursor concentrations are used) but rather by a too short reaction time. Then, larger coatings could only be achieved for longer reaction times, as depicted below.

For longer reaction times, the thickness of the solid coating approaches the one of the liquid coating measured with SMPS (Figure 8, right). For a TEOS concentration of 0.7 ppmv the thickness of the liquid coating measured before collection increases with time corroborating Equation (1). An induction period is reported here before condensation linearly increases with time, as for the condensation of organic vapors on metal nanoparticles [29]. The solid coating also increases with time. However, reaction times of a few minutes are required to convert all the liquid intermediates into the final solid coating with the closest approach of thicknesses measured with SMPS and TEM, respectively. Conversion of liquid intermediates into solid SiO_x coatings might still occur after collection as already reported [25]. Actually, much slower rates of conversion are expected here with post-DBD reaction temperatures close to ambient temperature, than in classical sol-gel process for the formation of oxides usually performed above a few hundred degrees Celsius [4].

Hence, the first gas phase reaction and the second condensation step leading to suspended droplets of intermediates are faster than the final self-reaction of silanol-like intermediates leading to the solid silica coating, for the near ambient temperature of the reaction. The step limiting the overall reaction rate is the conversion of the condensed intermediates into solid silica coating. The thickness of final solid SiO_xcoatings can be controlled from a nanometer to few tens of nanometers by playing on TEOS concentration for reaction times longer than 3 min, so that the SiO_x formation is nearly completed before collection, and on reaction times for TEOS concentrations in the ppmv range.

3.3.2. Influence of the Particle Material on the Coating Thickness

Different electrode materials (Pt, Cu, Ti, Ni, Fe) were used in the SDG producing different metal nanoparticles, sintered and coated without prior size classification. Contrary to the inert Pt, other metal nanoparticles were oxidized. A homogeneous SiO_x coating was created on spherical platinum, copper oxide and titanium oxide particles (Figure 9).

For the nickel and iron oxide particles, complete coalescence to spherical particles was not achieved so that the coating was performed with agglomerates. A very thin SiO_x coating was found on the outer surface of these particles, while most of the SiO_x was deposited inside the agglomerates. Therefore, the thicknesses of such heterogeneous coatings on nickel and iron oxide agglomerates have not been plotted here. The coating of agglomerates will be discussed in the next section.

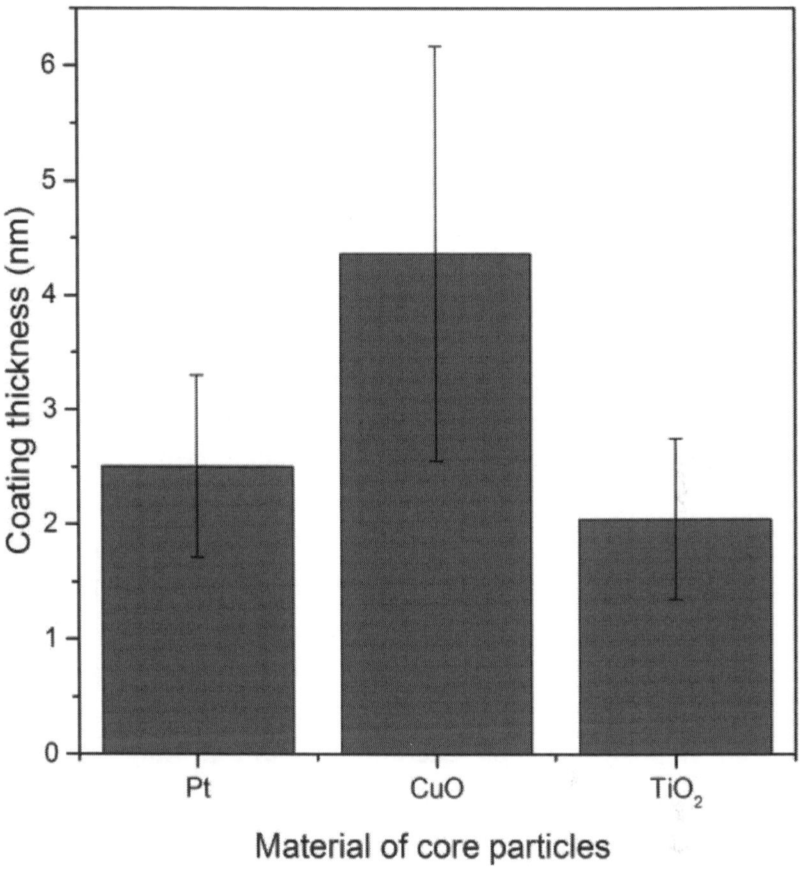

Figure 9. Influence of the particle material on the coating thickness from TEM analysis at a given TEOS concentration of 0.7 ppmv for$t_{Reaction}$ = 83 s.

3.3.3. Influence of the Particle Concentration on the Coating Thickness

For given TEOS concentration assumed to be related to the gas phase concentration of intermediates (N_{vap}), the coating thickness could also be controlled *versus* the total concentration of particles ($N_{part.}$) affecting the amount of condensable species per particle ($N_{vap}/N_{part.}$). The depletion effect could only be addressed for longer reaction times (>3 min) and smaller TEOS concentrations with a final forced condensation. Similar coating thickness was found even though the particle number concentrations differed by about two orders of magnitude for classified and non-classified Pt particles and for the different materials tested.

3.4. Heterogenous Coatings of Agglomerates

While spherical nanoparticles facilitate the investigation of the coating kinetics ("idealized system"), nanoparticles in technical applications are mostly encountered as agglomerates consisting of much smaller primary particles as shown in Figure 2 (left). The concave surface areas in the necks between individual primary particles as well as in the interior of the highly ramified agglomerates are most prone for condensational effects inducing capillary condensation [29]. Therefore, the coating of nanoparticle agglomerates is a superposition of structure-induced deposition and reaction to form SiO_x layers. Since deposition and coating inside the agglomerates will affect much less the agglomerate mobility compared to the coating of a sphere, the SMPS measurements can only provide a qualitative picture of the coating progress. Thus, another surface sensitive *in situ* technique is used here to follow the degree of surface coverage, *i.e.*, the aerosol photoemission (APE) [29]. Since SiO_2 has a high photothreshold (>10 eV [30]) that can only be overcome by extreme short wavelength UV light (far below UV-C), silica coatings are commonly employed to inhibit photocatalytic effects of sun blocking nanoparticles such as ZnO or TiO_2 [1] by suppressing the exchange of photogenerated species with the environment [2].

As starting point for the coating of Pt nanoparticle agglomerates the condensation of pure TEOS vapor without plasma (Figure

10, left) and with plasma turned on (Figure 10, right) was investigated as a function of the precursor concentration by tandem DMA measurements.

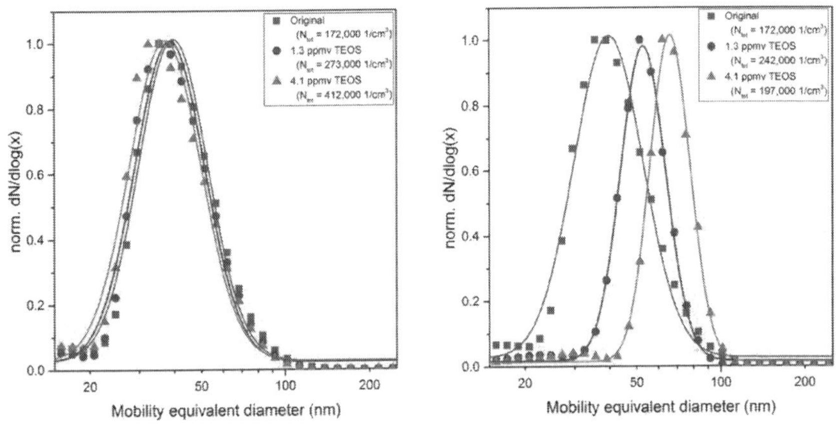

Figure 10. Change in mobility equivalent diameter depending on the precursor concentration without (**left**) and with (**right**) plasma for agglomerates t_{Reaction} = 83 s).

As shown in Figure 10 (left), small variations of the agglomerate size distribution were observed without a pronounced shift when the plasma was turned off. However, while with the plasma turned on a systematic shift of the size similar to the behavior of spherical particles was observed, the equivalent layer thickness appears to be smaller in comparison. On the one hand, the available surface area for the agglomerates is substantially higher (about a factor of 3) compared to the spherical Pt particles due to surface losses of the Pt particles as a consequence of coalescence but also due to particle losses in the sintering furnace. Therefore, the amount of condensing species per particle surface favors thicker layers on spherical particles. On the other hand, as outlined above, substantial condensation occurs inside agglomerates with a minor effect on the measured agglomerate mobility.

Next, the morphology of the solid silica coatings of the Pt agglomerates was studied with TEM for different TEOS concentrations and different reaction times. The coating in Figure 11 covers nearly the whole agglomerate. As already found for the spherical particles, the coating thickness of the agglomerates depends more on the reaction time than on the precursor concentration (*cf.* Figure 8).

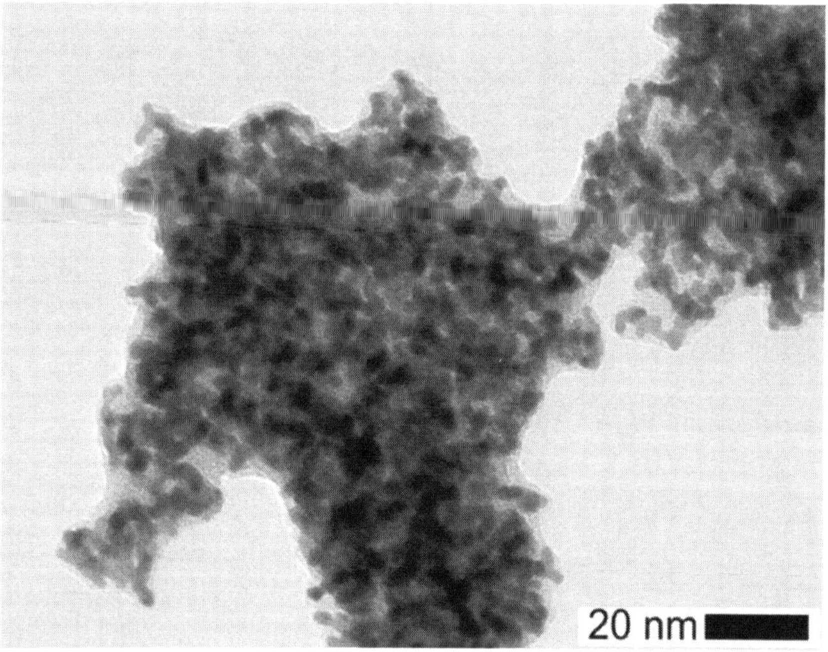

Figure 11. TEM micrograph of a coated Pt agglomerate after a reaction time of 185 s with a TEOS concentration of 0.7 ppmv showing a thick coating.

In order to follow the coating kinetics, photoemission measurements were done for Pt agglomerates as well as spheres as a function of the precursor concentration (Figure 12). The non-coated agglomerates ($c_{TEOS} = 0$) gave the highest signal corresponding to the largest amount of active surface area. Uncoated sintered spheres gave a smaller signal due to a reduced particle surface area, a consequence of particle losses (number concentration) and surface loss due to sintering. For the spheres, a small amount of TEOS is sufficient to obtain a complete coating, reflected in a rapid drop of the APE signal. Higher precursor concentrations increased the coating thickness, but even a small, hermetic coating was enough to prevent photoemission. Agglomerates, however, behaved differently in the way that coating was favored in the necks and the gaps between branches, leaving some outer particles exposed. The fraction of the uncoated primary particles contributing to the APE signal decreased with increasing TEOS concentration. At high TEOS concentrations the whole agglomerate was covered. Parallel to the complete coating of the agglomerates a restructuring of some branches to more compact structures was observed due to capillary forces. Such a compactization of nanoparticle agglomerates is well documented in the literature [31]. However, at lower precursor concentrations, restructuring did not occur. This observation raises the question if the mere coating of the necks would lead to mechanical stabilization against restructuring without hindering mass transport processes, which would enable interesting applications such as catalysis at higher temperatures.

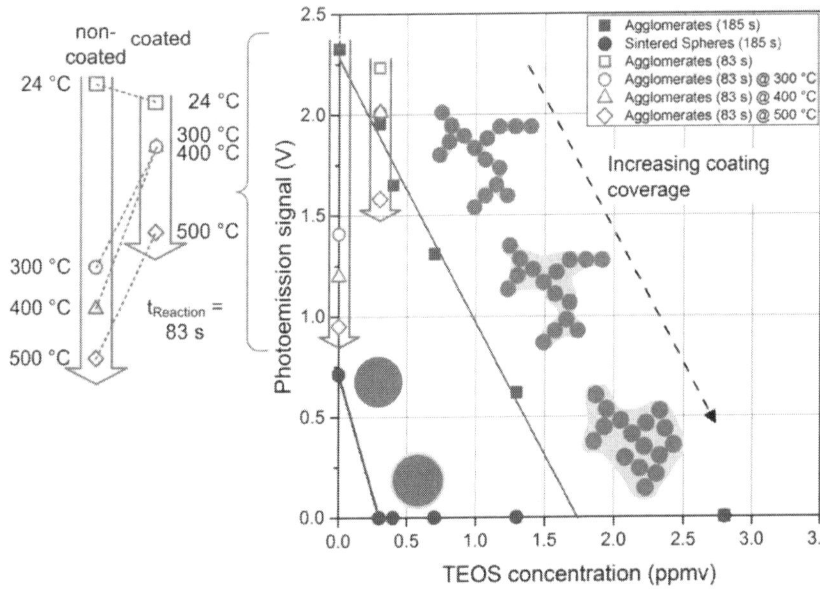

Figure 12. Measurement of the coating coverage by photoemission for Pt particles and sketches of the different kinds of observed coatings from TEM micrographs ($t_{Reaction}$ = 185 s). Additionally, the open, blue symbols show the change in APE signal depending on the sintering temperature, analogous to the particles shown inFigure 13 ($t_{Reaction}$ = 83 s).

Silica coatings have been employed in various forms to protect noble metal nanoparticle catalysts from sintering [32]. In this context, two opposing effects have to be managed: On the one hand, the thermal stability of the nanoparticles increases with increasing coating thickness. On the other hand, thicker silica layers hamper the mass transport across the coating reducing the catalytic activity. To overcome this problem, one solution is to obtain a sufficient porosity of the coating by high temperature treatment [1] or by adding pore building agents [32]. However, an easier and more versatile approach would be the precise control of the coating process by depositing silica only in the necks between the primary particles. Such a technique was applied by Kim and Ehrman [33] using TEOS as precursor to coat TiO_2 nanoparticles.

To test if such a precise control of the coating can also be realized in the post-plasma process investigated here minute amounts of TEOS precursor (0.3 ppmv) and short reaction times (83 s) were used. As a measure for the surface coverage, the APE signal of the coated Pt agglomerates was determined and compared to the uncoated agglomerates. As indicated in Figure 12 (open, blue symbols), the photoemission activity was only reduced by about 5% due to the silica coating (silica confirmed by EDX). It was verified by TEM microscopy that at these conditions most of the silica was in fact deposited in the necks between the primary particles leaving most of the Pt surface accessible for interaction with the surrounding gas molecules. In order to test if this coating was sufficient to stabilize the agglomerates against thermal restructuring, the sintering behavior of non-coated and coated particles was studied by passing the aerosol particles through a tube furnace at a constant gas flow rate either directly after the SDG or after coating. Figure 13 shows the TEM micrographs for furnace temperatures of 300, 400 and 500 °C, respectively. The applied coating, which conserved the surface functionality as shown above, was very thin and non-homogeneous. With increasing temperature, the differences in the morphologies of the coated and non-coated agglomerates became more pronounced. While the primary particles of the non-coated agglomerates visibly sintered to larger diameters even at temperatures as low as 300 °C, the coated particles showed much smaller changes and retained a branched structure. The thin coating did not completely prevent the sintering of primary particles but it preserved a higher amount of surface area. As indicated by the

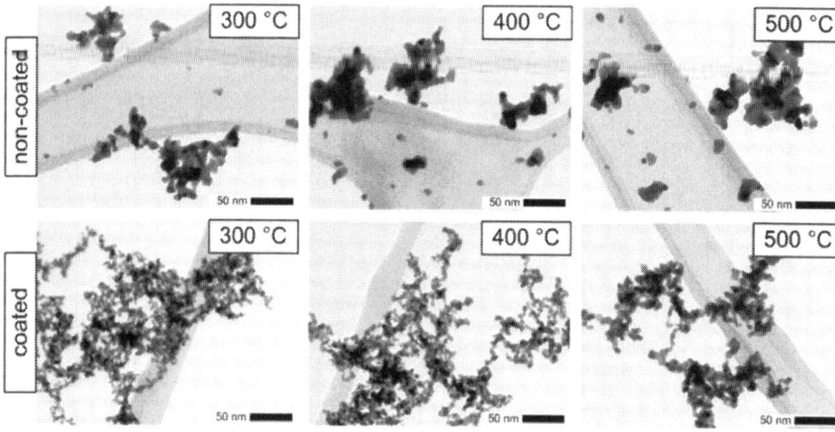

Figure 13. TEM micrographs showing the sintering behavior of non-coated (**top**) and coated (**bottom**) Pt particles during the transit through a tube furnace (residence time at 24 °C of 8 s).

blue symbols in Figure 12, also the photoactivity of the coated Pt agglomerates was only reduced by 12% for temperatures up to 400 °C and by 29% for a temperature of 500 °C. The uncoated agglomerates start with a slightly higher initial photoactivity at room temperature (24 °C), which decreases at elevated temperatures much faster compared to the coated particles (*cf.* insert in Figure 12). At 500 °C the remaining photoactivity of the coated particles is about 60% higher than the one of the uncoated Pt agglomerates. However, the coating process was not optimized so far and additional experiments will be necessary to explore the potential and to define the limits of this coating process for catalyst particles. Nevertheless, the first results of the post-plasma coating are promising and suggest further applications. Indeed, the operating conditions can be tuned to cover preferentially the interparticle contact zones between primary particles within agglomerates. This hampers the sintering of platinum agglomerates, retaining high apparent surface to much higher temperatures than usual sintering temperatures (e.g., for TiO_2 agglomerates that could be heated more than 400 °C above the usual sintering temperature [1]).

4. Conclusions

Homogeneous SiO_x coatings of spherical and agglomerated metal and metal oxide nanoparticles (Pt, CuO, and TiO_2) have been achieved in post-DBD. EDX results suggest some amount of carbon in the coating.

It has been shown that non-thermal plasma filaments induced in a dielectric barrier discharge (DBD) trigger post-DBD gas phase reactions of ozone and dinitrogen pentoxide with gaseous TEOS precursor, forming the final homogeneous SiO_x coatings.

This confirms the interest of post-DBD injection of the organo-silicon precursor to avoid electrode coating to achieve stable DBD production of active species with subsequent controlled thickness of SiO_xcoatings of nanoparticles *versus* TEOS concentration and reaction time.

The coating thickness can be controlled through the reaction time and the precursor concentration. At this point, it has to be underlined that larger coating thicknesses were measured before collection in the gas phase from size distribution measurements than after collection by TEM analysis. A possible explanation is the formation of a liquid intermediate product, while the subsequent reaction to a solid SiO_x coating requires more time and could still happen after collection.

The process seems independent of the particle material, so that very homogeneous coatings can be achieved on nanoparticles from different materials and with different structures (single spherical particles as well as agglomerates). Even thermally unstable materials could be coated at near ambient temperatures.

For agglomerates, the operating conditions can be tuned to cover preferentially the interparticle contact zones between primary particles. This hampers the sintering of platinum agglomerates, retaining high apparent surface to much higher temperatures than usual sintering temperatures. The coatings are suitable for enhanced thermal stability, as well as for tunable photoactivity.

Post-plasma chemistry is complex because of numerous possible pathways, and more analysis in different plasma and mixing conditions would be required to identify all reaction pathways. Then, the growth rate of the coating and the related coating thickness could be defined *versus* conditions of reaction (temperature and concentrations of reactants—plasma species and TEOS—as well as the relative concentration of intermediate condensable species and nanoparticles—$N_{\text{condensing vap}}/N_{\text{part}}$), probably also controlling the amount of liquid condensed per particle, and thus the final solid coating thickness.

Acknowledgments

The financial support by the German Research Foundation under grant WE 2331/18-1 is gratefully acknowledged.

Author Contributions

P.P. performed the experiments. N.J. and J.-P.B. characterized the plasma. P.P., J.-P.B. and A.P.W. conceived and designed the experiments, analyzed the data and wrote the paper.

Conflicts of Interest

The authors declare no conflict of interest.

References

1. Qi, F.; Moiseev, A.; Deubener, J.; Weber, A. Thermostable photocatalytically active TiO_2 anatase nanoparticles. *J. Nanopart. Res.* **2011**, *13*, 1325–1334.

2. Jansen, R.; Osterwalder, U.; Wang, S.Q.; Burnett, M.; Lim, H.W. Photoprotection: Part II. Sunscreen: Development, efficacy, and controversies. *J. Am. Acad. Dermatol.* **2013**, *69*.

3. Graf, C.; Vossen, D.L.J.; Imhof, A.; van Blaaderen, A. A General Method to Coat Colloidal Particles with Silica. *Langmuir* **2003**, *19*, 6693–6700.

4. Kodas, T.T.; Hampden-Smith, M.J. *Aerosol Processing of Materials*; Wiley-VCH: New York, NY, USA, 1999.

5. George, S.M. Atomic Layer Deposition: An Overview. *Chem. Rev.* **2010**, *110*, 111–131.

6. Teleki, A.; Heine, M.C.; Krumeich, F.; Akhtar, M.K.; Pratsinis, S.E. *In Situ* Coating of Flame-Made

TiO_2 Particles with Nanothin SiO_2 Films. *Langmuir* **2008**, *24*, 12553–12558.

7. Park, H.K.; Park, K.Y. Control of Particle Morphology and Size in Vapor-Phase Synthesis of Titania, Silica and Alumina Nanoparticles. *KONA Powder Part. J.* **2015**, *32*, 85–101.

8. Massines, F.; Sarra-Bournet, C.; Fanelli, F.; Naudé, N.; Gherardi, N. Atmospheric Pressure Low Temperature Direct Plasma Technology: Status and Challenges for Thin Film Deposition. *Plasma Process. Polym.* **2012**, *9*, 1041–1073.

9. King, D.M.; Liang, X.; Burton, B.B.; Kamal Akhtar, M.; Weimer, A.W. Passivation of pigment-grade TiO_2 particles by nanothick atomic layer deposited SiO_2 films. *Nanotechnology* **2008**, *19*.

10. Boies, A.M.; Calder, S.; Agarwal, P.; Lei, P.; Girshick, S.L. Chemical Kinetics of Photoinduced Chemical Vapor Deposition: Silica Coating of Gas-Phase Nanoparticles. *J. Phys. Chem. C* **2012**, *116*, 104–114.

11. Jiménez, C.; De Barros, D.; Darraz, A.; Deschanvres, J.-L.; Rapenne, L.; Chaudouët, P.; Méndez, J.E.; Weiss, F.; Thomachot, M.; Sindzingre, T.; *et al.* Deposition of TiO_2 thin films by atmospheric plasma post-discharge assisted injection MOCVD. *Surf. Coat. Technol.* **2007**, *201*, 8971–8975.

12. Borra, J.-P.; Valt, A.; Arefi-Khonsari, F.; Tatoulian, M. Atmospheric Pressure Deposition of Thin Functional Coatings: Polymer Surface Patterning by DBD and Post-Discharge Polymerization of Liquid Vinyl Monomer from

Surface Radicals. *Plasma Process. Polym.* **2012**, *9*, 1104–1115.

13. Fanelli, F.; Fracassi, F. Aerosol-Assisted Atmospheric Pressure Cold Plasma Deposition of Organic–Inorganic Nanocomposite Coatings. *Plasma Chem. Plasma Process.* **2014**, *34*, 473–487.

14. Mori, T.; Tanaka, K.; Inomata, T.; Takeda, A.; Kogoma, M. Development of silica coating methods for powdered pigments with atmospheric pressure glow plasma. *Thin Solid Films* **1998**, *316*, 89–92.

15. Kogoma, M.; Tanaka, K.; Takeda, A. Powder Treatments using Atmospheric Pressure Glow Plasma (Silica Coating of TiO_2 Fine Powder). *J. Photopolym. Sci. Technol.* **2005**, *18*, 277–280.

16. Brueser, V.; Hahnel, M.; Kersten, H. Thin Film Deposition on Powder Surfaces Using Atmospheric Pressure Discharge. In *New Vistas in Dusty Plasmas: Fourth International Conference on the Physics of Dusty PlasmasAIP Conference Proceedings*; American Institute of Physics: College Park, MD, USA, 2005; Volume 799, pp. 343–346.

17. Vons, V.; Creyghton, Y.; Schmidt-Ott, A. Nanoparticle production using atmospheric pressure cold plasma. *J. Nanopart. Res.* **2006**, *8*, 721–728.

18. Nessim, C.; Boulos, M.; Kogelschatz, U. In-flight coating of nanoparticles in atmospheric-pressure DBD torch plasmas. *Eur. Phys. J. Appl. Phys.* **2009**, *47*.

19. Meuller, B.O.; Messing, M.E.; Engberg, D.L.J.; Jansson, A.M.; Johansson, L.I.M.; Norlén, S.M.; Tureson, N.; Deppert, K. Review of Spark Discharge Generators for Production of Nanoparticle Aerosols. *Aerosol Sci. Technol.* **2012**, *46*, 1256–1270.

20. Borra, J.-P.; Jidenko, N.; Hou, J.; Weber, A. Vaporization of bulk metals into single-digit nanoparticles by non-thermal plasma filaments in atmospheric pressure dielectric barrier discharges. *J. Aerosol Sci.* **2015**, *79*, 109–125.

21. Weber, A.P.; Seipenbusch, M.; Kasper, G. Application of Aerosol Techniques to Study the Catalytic Formation of Methane on Gasborne Nickel Nanoparticles. *J. Phys. Chem. A* **2001**, *105*, 8958–8963.

22. Borra, J.-P. Nucleation and aerosol processing in atmospheric pressure electrical discharges: Powders production, coatings and filtration. *J. Phys. Appl. Phys.* **2006**, *39*, R19–R54.

23. Okuyama, K.; Fujimoto, T.; Hayashi, T.; Adachi, M. Gas-phase nucleation in the tetraethylorthosilicate (TEOS)/O_3 APCVD process. *AIChE J.* **1997**, *43*, 2688–2697.

24. Fujino, K.; Nishimoto, Y.; Tokumasu, N.; Maeda, K. Silicon Dioxide Deposition by Atmospheric Pressure and Low-Temperature CVD Using TEOS and Ozone. *J. Electrochem. Soc.* **1990**, *137*, 2883–2887.

25. Romet, S.; Couturier, M.F.; Whidden, T.K. Modeling of silicon dioxide chemical vapor deposition from

tetraethoxysilane and ozone. *J. Electrochem. Soc.* **2001**, *148*, G82–G90.

26. Bourgeois, E.; Jidenko, N.; Alonso, M.; Borra, J.P. DBD as a post-discharge bipolar ions source and selective ion-induced nucleation *versus* ions polarity. *J. Phys. Appl. Phys.* **2009**, *42*.

27. Mathon, R.; Jidenko, N.; Borra, J.-P. Characterization of a Post-DBD Aerosol Bipolar Diffusion Neutralizer for SMPS Size Distribution Measurements. In Proceedings of 2016 European Aerosol Conference, Tours, France, 4–6 September 2016.

28. Hinds, W.C. *Aerosol Technology: Properties, Behavior, and Measurement of Airborne Particles*, 2nd ed.; Wiley: New York, NY, USA, 1999.

29. Poostforooshan, J.; Rennecke, S.; Gensch, M.; Beuermann, S.; Brunotte, G.-P.; Ziegmann, G.; Weber, A.P. Aerosol Process for the *In Situ* Coating of Nanoparticles with a Polymer Shell. *Aerosol Sci. Technol.* **2014**, *48*, 1111–1122.

30. Wellert, S. Wechselwirkung von Elektronen und Molekülen Mit Einzelnen SiO_2-Nanopartikeln: Massenanalyse in Einer Vierpolfalle. Ph.D. Thesis, Fakultät für Naturwissenschaften, Technische Universität Chemnitz, Chemnitz, Germany, 2003.

31. Miljevic, B.; Surawski, N.C.; Bostrom, T.; Ristovski, Z.D. Restructuring of carbonaceous particles upon exposure to organic and water vapours. *J. Aerosol Sci.* **2012**, *47*, 48–57.

32. Park, J.-N.; Forman, A.J.; Tang, W.; Cheng, J.; Hu, Y.-S.; Lin, H.; McFarland, E.W. Highly Active and Sinter-Resistant Pd-Nanoparticle Catalysts Encapsulated in Silica. *Small* **2008**, *4*, 1694–1697.

33. Kim, S.; Ehrman, S.H. Capillary Condensation onto Titania (TiO_2) Nanoparticle Agglomerates.*Langmuir* **2007**, *23*, 2497–2504.

6

CHAPTER

Synthesis of Lithium Metal Oxide Nanoparticles by Induction Thermal Plasmas

Manabu Tanaka [1], Takuya Kageyama [2], Hirotaka Sone [2], Shuhei Yoshida [3], Daisuke Okamoto [2] and Takayuki Watanabe [1,*]

[1]*Department of Chemical Engineering, Faculty of Engineering, Kyushu University, 744 Motooka, Nishi-ku, Fukuoka 819-0395, Japan*

[2]*Department of Chemical Systems and Engineering, Graduate School of Engineering, Kyushu University, 744 Motooka, Nishi-ku, Fukuoka 819-0395, Japan*

[3]*Chemical Engineering Course, School of Engineering, Kyushu University, 744 Motooka, Nishi-ku, Fukuoka 819-0395, Japan*

Abstract

Lithium metal oxide nanoparticles were synthesized by induction thermal plasma. Four different systems—Li–Mn, Li–Cr, Li–Co, and Li–Ni—were compared to understand formation mechanism of Li–Me oxide nanoparticles in thermal plasma process. Analyses of X-ray diffractometry and electron microscopy showed that Li–Me oxide nanoparticles were successfully synthesized in Li–Mn, Li–Cr, and Li–Co systems. Spinel structured $LiMn_2O_4$ with truncated octahedral shape was formed. Layer structured $LiCrO_2$ or $LiCoO_2$ nanoparticles with polyhedral shapes were also synthesized in Li–Cr or Li–Co systems. By contrast, Li–Ni oxide nanoparticles were not synthesized in the Li–Ni system. Nucleation temperatures of each metal in the considered system were evaluated. The relationship between the nucleation temperature and melting and boiling points suggests that the melting points of metal oxides have a strong influence on the formation of lithium metal oxide nanoparticles. A lower melting temperature leads to a longer reaction time, resulting in a higher fraction of the lithium metal oxide nanoparticles in the prepared nanoparticles.

Keywords: thermal plasmas; lithium metal oxide; nanoparticle formation mechanism

1. Introduction

Nanoparticles have become widely utilized due to their enhanced and unique properties relative to bulk materials. The preparation of nanoparticles can be classified into physical and chemical methods. The physical methods include mechanical milling [1,2], laser ablation [3,4,5], and other aerosol processes with energy sources to provide a high temperature. Among these methods, attractive material processing with thermal plasmas have been proposed for the nanoparticles production. This is because thermal plasmas offer unique advantages; high enthalpy to enhance reaction kinetics, high chemical reactivity, rapid quenching rate in the range of 10^{3-6} K/s, and selectivity of atmosphere in accordance with the required chemical reactions. Thermal plasmas are capable of evaporating large amount of raw materials, even with high melting and boiling temperatures [6,7,8,9]. Furthermore, high-purity nanoparticles can be synthesized in an induction thermal plasma because thermal plasma can be generated in a plasma torch without internal electrodes [10,11]. These advantages of thermal plasmas have brought about advances in plasma chemistry and plasma processing [12,13,14,15].

Lithium metal oxides have attracted many researchers because of their unique properties as cathode for lithium-ion batteries [16,17,18], CO_2 sorption material [19], and other magnetic, electrochemical materials. Layer structured $LiCoO_2$ is widely employed as cathodes in commercial battery applications, in spite of its toxicity and high cost. To solve the economic and environmental problems, alternatives of $LiCoO_2$ have been intensively explored. One of the alternatives is spinel-structured $LiMn_2O_4$, which provides a promising high-voltage cathode material for lithium-ion batteries owing to their high theoretical energy density, low cost, and eco-friendliness [18,20,21,22]. $LiNiO_2$ is also a candidate due to its excellent cycle life, with negligible capacity fading *etc.* [17,23]. The liquid phase method is generally used in the synthesis of lithium metal oxide nanoparticles; however, productivity of the nanoparticles in the liquid phase method is insufficient for industrial application. Therefore, the synthesis method of lithium oxide nanoparticles with a high productivity is strongly demanded.

Thermal plasmas are expected to be promising energy sources to fabricate nanoparticles at high productivity from micron-sized powder as raw materials. Here, one-step synthesis of lithium-metal oxide nanoparticles with induction thermal plasma is studied. The purpose of the present study is to synthesize the lithium metal oxide nanoparticles via the induction thermal plasma and to investigate the formation mechanism of lithium-metal oxide nanoparticles. Different lithium metal systems—Li–Mn, Li–Cr, Li–Co, and Li–Ni—were compared to understand the formation mechanism.

2. Results and Discussion

2.1. Experimental Results

Figure 1 shows X-ray diffraction spectra of the prepared nanoparticles by the induction thermal plasma in different systems—Li–Mn, Li–Co, Li–Cr, and Li–Ni. In the case of the Li–Mn system, main diffraction peaks correspond to spinel-structured $LiMn_2O_4$, while diffraction peaks of Mn_3O_4 are also found. In cases of Li–Co and Li–Cr, layer-structured $LiCoO_2$ and $LiCrO_2$ were found as well as their oxides. In the Li–Ni case, $Li_{0.4}Ni_{1.6}O_2$ and unreacted Li_2CO_3 were confirmed. These results clearly show the strong influence of the constituent metals on the formation of lithium oxide nanoparticles. Moreover,Figure 2 indicates composition of the prepared nanoparticles analyzed from X-ray diffractometry (XRD) spectra. These compositions of the prepared nanoparticles were used to evaluate the ratio of metal that reacted with Li. Evaluated reaction ratio will be discussed in the following section.

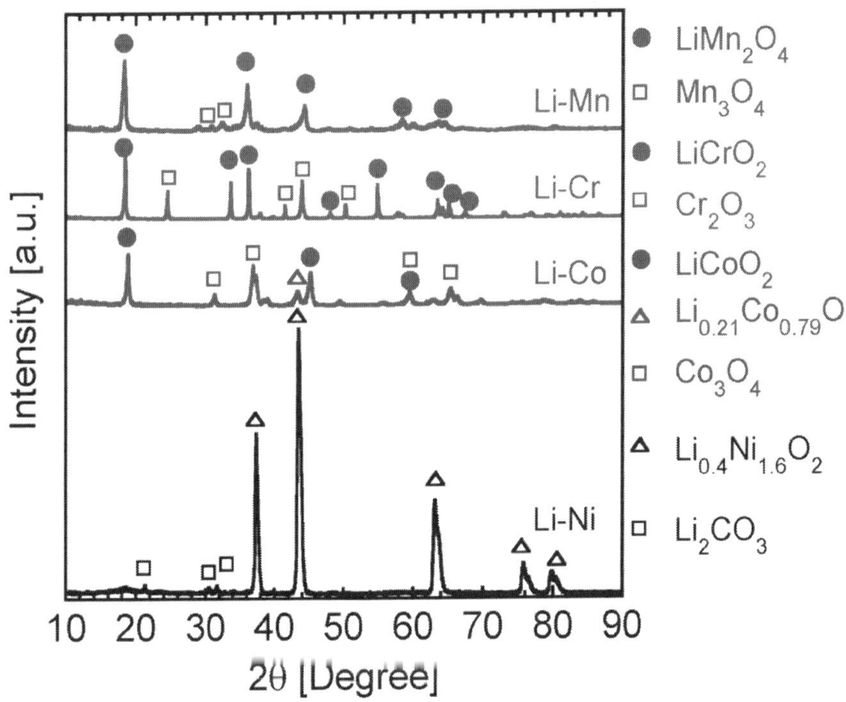

Figure 1. X-ray diffractometry (XRD) spectra of as-prepared nanoparticles by induction thermal plasma in different systems, Li–Mn, Li–Co, Li–Cr, and Li–Ni.

Morphologies and particle sizes of the prepared nanoparticles were observed via transmission electron microscopy (TEM). Figure 3 shows the representative TEM images and the particle size distributions of the nanoparticles in different systems. Many spherical particles with 60 nm in mean diameter are observed in the Li–Ni system. In the cases of Li–Mn, Li–Co, and Li–Cr, most of the particles with 50–80 nm in mean diameters have polyhedral shapes including quadrangular, pentagonal, and hexagonal shapes. In particular, many particles with a hexagonal shape can be found as shown inFigure 3a. Hence, the spinel-structured $LiMn_2O_4$ is considered to have a hexagonal shape because $LiMn_2O_4$ is a major product in the Li–Mn system, according to XRD results.

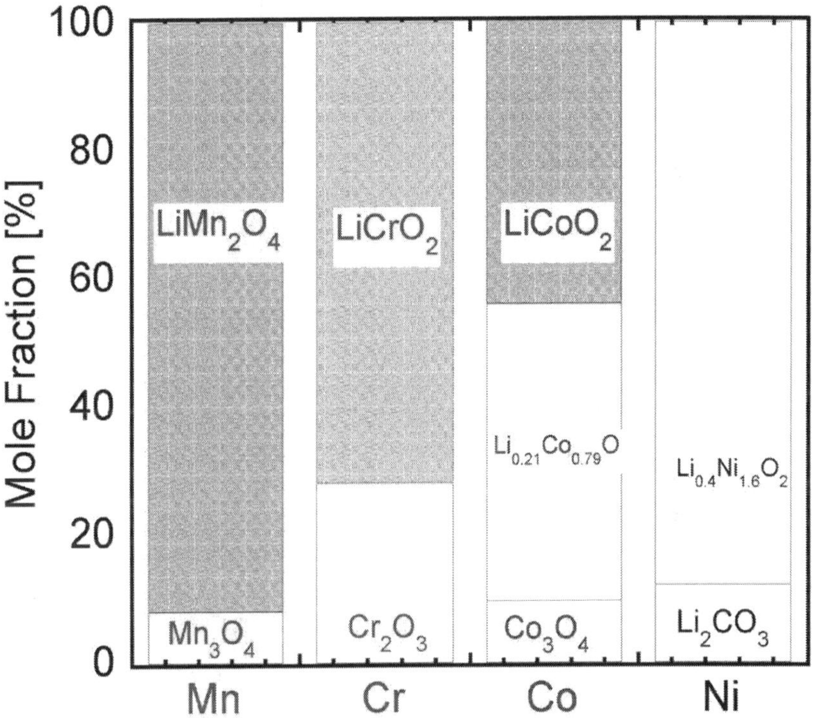

Figure 2. Composition of prepared nanoparticles by induction thermal plasma in different systems, Li–Mn, Li–Co, Li–Cr, and Li–Ni.

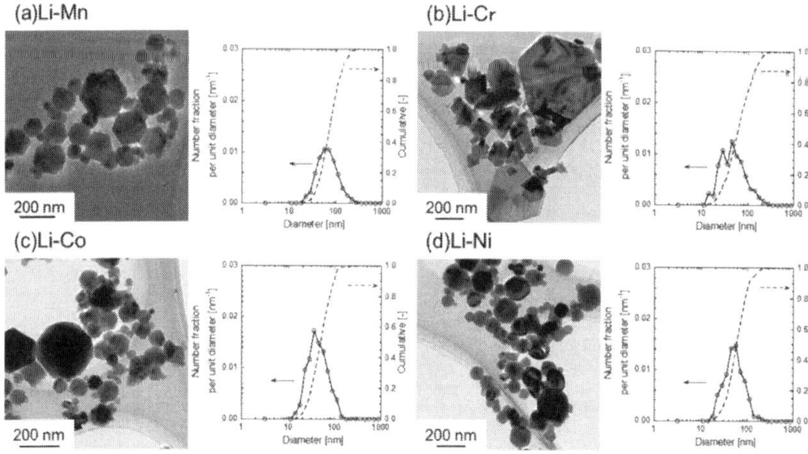

Figure 3. (**a**) Representative transmission electron microscopy (TEM) image and particle size distribution of prepared nanoparticles by induction thermal plasma in Li–Mn system; (**b**) that in Li–Cr system; (**c**) that in Li–Co system; (**d**) that in Li–Ni system.

Scanning electron microscopy (SEM) observation was carried out to clarify morphology of $LiMn_2O_4$more specifically than TEM observation. Figure 4 shows the representative SEM image of the prepared nanoparticles in the Li–Mn system. Many particles have truncated an octahedral shape as shown inFigure 4. Previous research on morphology of ferrite nanoparticles synthesized by the induction thermal plasma reported that the spinel-structured nanoparticles had a truncated octahedral shape [24]. The spinel-structured nanoparticles synthesized in thermal plasma have a truncated octahedral shape, although the stable structure of the spinel-structured particles are generally considered to be of an octahedral shape. The reason for this truncation is currently under investigation.

Figure 4. Representative scanning electron microscopy (SEM) image of prepared nanoparticles in Li–Mn system.

2.2. Nanoparticle Formation Mechanism

Homogeneous nucleation temperatures of metals considered in the present study were estimated based on nucleation theory considering non-dimensional surface tension [25]. The homogeneous nucleation rate J can be expressed as

$$J = \frac{\beta_{ij} n_s^2 S}{12} \sqrt{\frac{\Theta}{2\pi}} \exp\left(\Theta - \frac{4\Theta^3}{27(\ln S)^2}\right) \tag{1}$$

where S is the saturation ratio and n_s is the equilibrium saturation monomer concentration at temperature T. β is the collision frequency function. The dimensionless surface tension is given by the following equation:

$$\Theta = \frac{\sigma s_1}{kT} \tag{2}$$

where σ is the surface tension and s_1 is the monomer surface area. The surface tension and the saturation ratio have a dominant influence on determining the nucleation rate. Stable nuclei are observed experimentally when the nucleation rate is over 1.0 $cm^{-3}\cdot s^{-1}$. Hence, the corresponding value of the saturation ratio is defined as the critical saturation ratio. The detailed procedure to estimate the nucleation rate and corresponding nucleation temperature can be found in previous works [10,25].

The relationship between the calculated nucleation temperature and the boiling and melting points is summarized in Figure 5. Because of the unknown properties of metal oxides, only melting point oxides are plotted for metal oxides. These temperatures indicate that the melting points of the oxides are higher than the nucleation temperatures of pure metals in each Li–Me system. Therefore, nucleation of metal oxides is considered to occur at first. Li oxide and metal vapors co-condense onto the nuclei just after the nucleation starts.

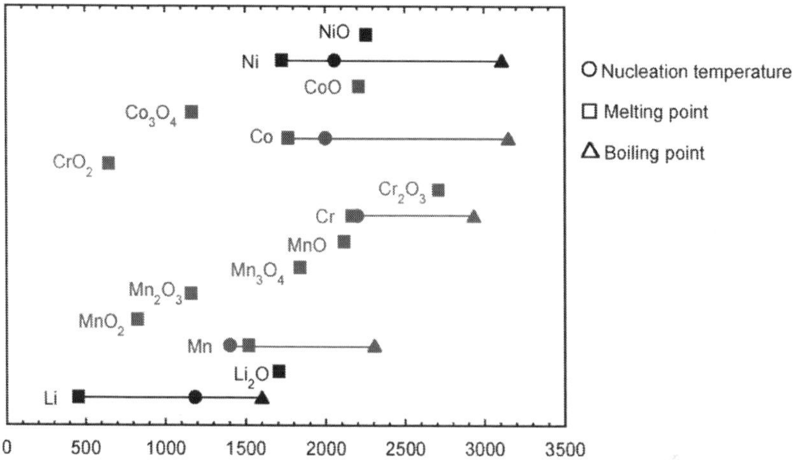

Figure 5. Relationship between nucleation temperature and melting and boiling points of considered metals and metal oxides.

The above mechanism can be proposed as a common mechanism for all considered Li–Me systems because the relationship between the nucleation temperature and the melting and boiling temperature of their oxides shows the same trend. However, experimental results show a different ratio of metal reactive with Li to total metal. In the cases of Li–Cr and Li–Mn, high ratios of the reactive metals were obtained, while a low ratio was obtained in Li–Ni system. Then, melting points of each oxide in different system were focused because the reaction rate of metal oxide particles and the condensed lithium would drastically decrease after complete solidification of the growing particles.

Figure 6 shows the relationship between the lowest melting temperature of metal oxides for each Li–Me system and the reaction ratios. The results suggest a lower melting point of oxide leads to a higher reaction ratio. This can be explained by the different reaction time during the nanoparticle formation process. Figure 7 summarizes the formation mechanism of Li–Me oxide nanoparticles. A lower melting point of oxide leads to a longer residence time of growing particles in a liquid-like state, resulting in the longer reaction time with condensed lithium oxide. Consequently, a higher reaction ratio can be obtained in the Li–Me system, as in that of Li–Mn and Li–Cr. On the other hand, a shorter residence time of the growing particles in a liquid-like state leads to a shorter reaction time, resulting in a lower reaction ratio. These results suggest that the melting point of metal oxide has a strong influence on the reaction with lithium oxide.

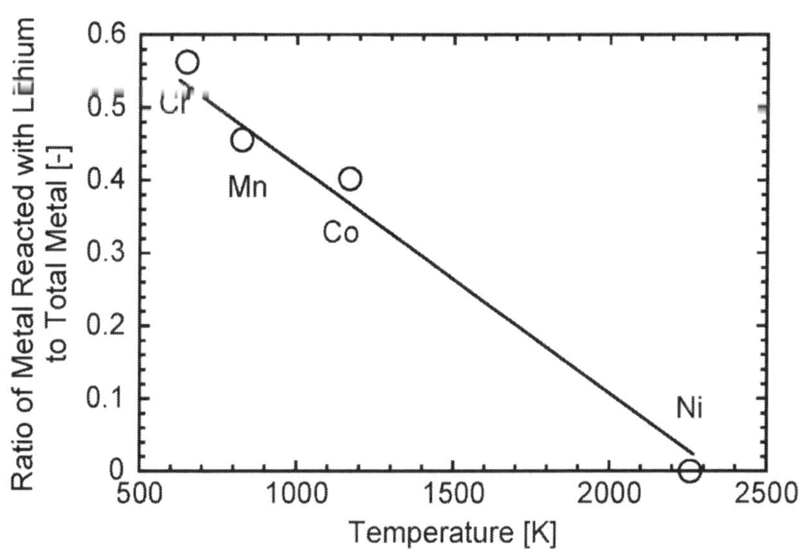

Figure 6. Relationship between lowest melting points of metal oxides and the reaction ratios for each Li–Me system.

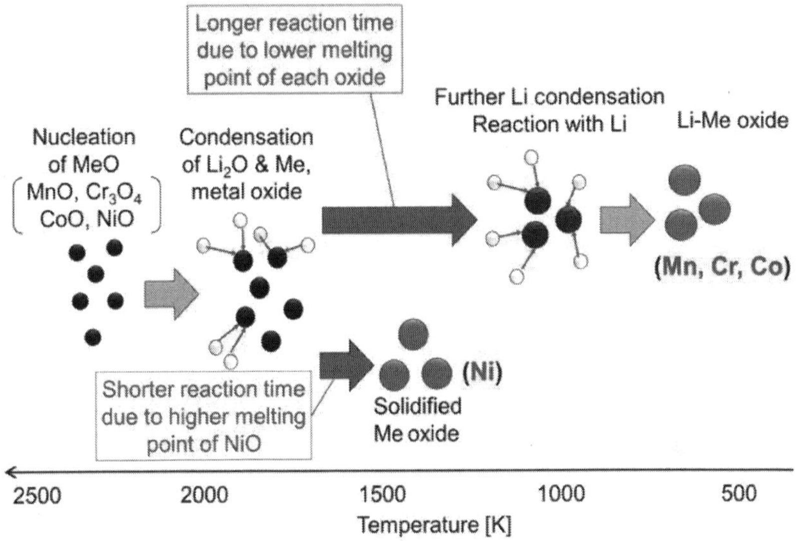

Figure 7. Conceptual diagram of formation mechanism of Li–Me oxide nanoparticles in induction thermal plasma.

The above mechanism can explain the obtained results well, although this is still only a hypothesis. Further experimental and theoretical investigation will be required to understand the formation mechanism of complicated oxide nanoparticles in thermal plasmas.

3. Experimental Section

3.1. Experimental Setup and Conditions

A schematic illustration of the experimental setup for Li–Me oxide nanoparticle fabrication is presented in Figure 8a. This equipment is composed of the plasma torch, the reaction chamber where the nanoparticles are synthesized, and the filter unit. Figure 8b shows the enlarged illustration of the plasma torch. High temperature plasma of more than 10,000 K was generated in the plasma torch by induction heating at 4 MHz. The input power was controlled at 20 kW. Ar was used for the carrier gas of the raw powder at 3 L/min and inner gas at 5 L/min. A mixture of argon and oxygen were used for the plasma forming gas at 60 L/min. Mixture of Li_2CO_3 with 3.5 μm in diameter and metal or metal oxide with 3–10 μm in mean diameter were injected from the powder feeder by Ar carrier gas. Different metals including Mn, Cr, Co, and Ni were compared to investigate the formation mechanism of Li–Me (Mn, Cr, Co, and Ni) nanoparticles. The composition ratio of Li_2CO_3 to metal was 0.5. The powder feed rate was fixed at 400 mg/min. These experimental conditions are listed in Table 1.

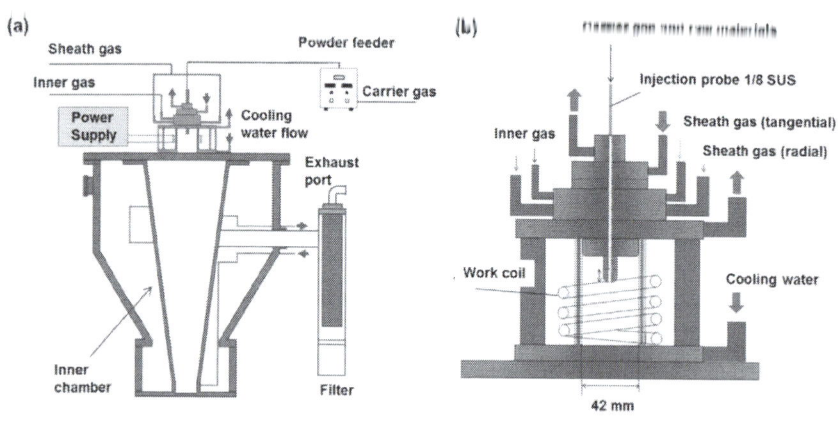

Figure 8. Schematic illustration of induction thermal plasma system for nanoparticles fabrication (**a**) and enlarged illustration of plasma torch (**b**).

Table 1. Typical experimental conditions.

Plasma Conditions				
Input power		20 kW		
Frequency		4 MHz		
Pressure		101.3 kPa		
Sheath gas	Ar: 57.5 L/min		O_2: 2.5 L/min	
Inner gas		Ar: 5 L/min		
Carrier gas		Ar: 3 L/min		
Discharge time		5 min		
Feed rate		400 mg/min		
Raw Materials				
System	Li–Mn	Li–Cr	Li–Co	Li–Ni
Raw powders	Li_2CO_3, MnO_2	Li_2CO_3, Cr	Li_2CO_3, Co	Li_2CO_3, Ni
Li/Me ratio		0.5		

Prepared nanoparticles were collected from the filter and the inner wall of the reaction chamber. Unevaporated raw materials were not confirmed according to SEM observation of the collected particles. This fact implies that the fed raw materials were completely evaporated in the high temperature region of the thermal plasma and converted into nanoparticles during the quenching process. Therefore, the mole fractions of the Li–Me oxides indicated in Figure 2 correspond to the yields of the Li–Me oxide in this nanoparticle fabrication process by the induction thermal plasma.

3.2. Characterization of Prepared Nanoparticles

The phase identification of the prepared nanoparticles was determined by X-ray diffractometry (XRD, Multiflex, Rigaku Co., Tokyo, Japan), operated with Cu Kα source (λ = 0.1541 nm). The diffraction data was collected using a continuous scan mode with a speed of 2 degree/min in the region of 10–90 degrees with a step size of 0.04 degrees. The accelerating voltage and applied current was 40 kV and 50 mA, respectively. The quantitative analysis of the composition of the prepared nanoparticles was conducted based on the whole-powder-pattern-decomposition (WPPD) method with the assumption that no amorphous particles were included in the prepared nanoparticles.

The particle morphology and size distribution of the prepared nanoparticles were observed by TEM (JEM-2100HCKM, JEOL Ltd., Tokyo, Japan), operated at an accelerating voltage of 200 kV. The TEM specimens were prepared by dispersing the as-prepared nanoparticles in ethanol and placing a few drops of the dispersion on a carbon-grid. Furthermore, the 3D particle morphology was observed by field emission (FE)-SEM (SII TES+ Zeiss ULTRA55, Carl Zeiss, Oberkochen, Germany).

4. Conclusions

Lithium metal oxide nanoparticles were synthesized in induction thermal plasma and formation mechanism was investigated. Obtained remarks are as follows:

a) Lithium metal oxide nanoparticles were synthesized in different Li–Me (Mn, Cr, Co, and Ni) systems. In the case of Li–Mn, Li–Cr, and Li–Co, lithium-metal oxide nanoparticles were successfully synthesized, while Li–Ni oxides were not synthesized in the Li–Ni system.

b) The spinel-structured $LiMn_2O_4$ with a truncated octahedral shape was synthesized in Li–Mn system, although the stable shape of the spinel structure was an octahedral shape.

c) The relationship between nucleation temperature and boiling and melting points of the considered metals and their oxides suggests the following formation mechanism: Metal oxide starts to nucleate at first. Then, vapors of metal and lithium oxide co-condense on the metal nuclei with an oxidation reaction.

d) Melting point of metal oxides is an important factor in determining the final product of the Li–Me composite. A lower melting point of metal oxide leads to a longer reaction time, resulting in higher yields of the Li–Me composite.

e) Nanomaterial fabrication with induction thermal plasma enables the production of high-purity nanoparticles of Li–Me oxide at high productivity.

Acknowledgments

The authors thank the Ultramicroscopy Research Center, Kyushu University for TEM and FE-SEM observation. The authors also thank the Center of Advanced Instrumental Analysis, Kyushu University for XRD analyses.

Author Contributions

Manabu Tanaka, Hirotaka Sone and Takayuki Watanabe coordinated the study. Manabu Tanaka, Takuya Kageyama, Shuhei Yoshida and Daisuke Okamoto conducted the experiments. Takuya Kageyama, Hirotaka Sone and Shuhei Yoshida conducted analyses of the produced nanoparticles. Takuya Kageyama conducted the data analyses. Manabu Tanaka and Takayuki Watanabe prepared the manuscript. All authors read and approved the manuscript.

Conflicts of Interest

The authors declare no conflict of interest.

References

1. Janot, R.; Guerard, D. One-step synthesis of maghemite nanometric powders by ball-milling. *J. Alloys Compd.* **2002**, *333*, 302–307.

2. Pithawalla, Y.B.; El Shall, M.S.; Deevi, S.C. Synthesis and characterization of nanocrystalline iron aluminide particles. *Intermetallics* **2000**, *8*, 1225–1231.

3. Peterson, S.; Barickowski, S. *In situ* bioconjugation: Single step approach to tailored nanoparticle-bioconjugates by ultrashort pulsed laser ablation. *Adv. Funct. Mater.* **2009**, *19*, 1167–1172.

4. Mafune, F.; Khono, J.Y.; Takeda, Y.; Kondow, T. Formation of gold nanoparticles by laser ablation in aqueous solution of surfactant. *J. Phys. Chem. B* **2001**, *105*, 5114–5120.

5. Yasukuni, R.; Horinaka, T.; Asahi, T. Preparation of perylenediimide nanoparticle colloids by laser ablation in water and their optical properties. *Jpn. J. Appl. Phys.* **2010**, *49*.

6. Shigeta, M.; Watanabe, T. Numerical analysis for co-condensation processes in silicide nanoparticle synthesis using induction thermal plasmas at atmospheric pressure conditions. *J. Mater. Res.* **2005**, *20*, 2801–2811.

7. Shigeta, M.; Watanabe, T. Growth mechanism of silicon-based functional nanoparticles fabricated by inductively coupled thermal plasmas. *J. Phys. D* **2007**, *40*, 2407–2419.

8. Shigeta, M.; Watanabe, T. Numerical investigation of cooling effect on platinum nanoparticle formation in inductively coupled thermal plasmas. *J. Appl. Phys.* **2008**, *103*.

9. Shigeta, M.; Murphy, A.B. Thermal plasmas for nanofabrication. *J. Phys. D* **2011**, *44*.

10. Tanaka, M.; Noda, J.; Watanabe, T.; Matsuno, J.; Tsuchiyama, A. Formation mechanism of metal embedded amorphous silicate nanoparticles by induction thermal plasmas. *J. Phys. Conf. Ser.* **2014**, *518*.

11. Cheng, Y.; Tanaka, M.; Watanabe, T.; Choi, S.-Y.; Shin, M.-S.; Lee, K.-H. Synthesis of Ni_2B nanoparticles by RF thermal plasma for fuel cell catalyst. *J. Phys. Conf. Ser.* **2014**, *518*.

12. Shigeta, M.; Watanabe, T. Effect of precursor fraction on silicide nanopowder growth under a thermal plasma condition: A computational study. *Powder Technol.* **2015**, *288*, 191–201.

13. Watanabe, T.; Liu, Y.; Tanaka, M. Investigation of electrode phenomena in an innovative thermal plasma for glass melting. *Plasma Chem. Plasma Proc.* **2014**, *34*, 443–456.

14. Liang, F.; Tanaka, M.; Watanabe, T. Measurement of anode surface temperature in carbon nanomaterial

production by arc discharge method. *Mater. Res. Bull.* **2014**, *60*, 158–165.

15. Tanaka, M.; Watanabe, T. Enhanced vaporization from molten metal surface by argon-hydrogen arc plasma. *Jpn. J. Appl. Phys.* **2013**, *52*.

16. Xiao, X.; Wang, L.; Wang, D.; He, Q.; Peng, Q.; Li, Y. Hydrothermal synthesis of orthorhombic $LiMnO_2$ nano-particles and $LiMnO_2$ nanorods and comparison of their electrochemical performances. *Nano. Res.* **2009**, *2*, 923–930.

17. Kalyani, P.; Kalaiselvi, N. Various aspects of $LiNiO_2$ chemistry: A review. *Sci. Technol. Adv. Mater.* **2005**, *6*, 689–703.

18. Curtis, C.J.; Wang, J.; Schulz, D.L. Preparation and characterization of $LiMn_2O_4$ spinel nanoparticles as cathode materials in secondary Li batteries. *J. Electrochem. Soc.* **2004**, *151*, A590–A598.

19. Ida, J.; Lin, Y.S. Mechanism of high-temperature CO_2 sorption on lithium zirconate. *Environ. Sci. Technol.* **2003**, *37*, 1999–2004.

20. Shaju, K.M.; Bruce, P.G. Macroporous $Li(Ni_{1/3}Co_{1/3}Mn_{1/3})O_2$: A high power and high energy cathode for rechargeable lithium batteries. *Adv. Mater.* **2006**, *18*.

21. Canulescu, S.; Papadopoulou, E.L.; Anglos, D.; Lippert, T.; Schneider, W.; Wokaun, A. Mechanism of the laser plume expansion during the ablation of $LiMn_2O_4$. *J. Appl. Phys.* **2009**, *105*.

22. Kim, D.K.; Muralidharan, P.; Lee, H.-W.; Ruffo, R.; Yang, Y.; Chan, C.K.; Peng, H.; Huggins, R.A.; Cui, Y. Spinel $LiMn_2O_4$ nanorods as lithium ion battery cathodes. *Nano Lett.* **2008**, *8*, 3948–3952.

23. Ohzuku, T.; Ueda, A.; Nagayama, M. Electrochemistry and structural chemistry of $LiNiO_2$ (R3m) for 4 volt secondary lithium cells. *J. Electrochem. Soc.* **1993**, *140*, 1862–1870.

24. Swaminathan, R.; Willard, M.A.; McHenry, M.E. Experimental observations and nucleation and growth theory of polyhedral magnetic ferrite nanoparticles synthesized using an RF plasma torch. *Acta Mater.* **2006**, *54*, 807–816.

25. Girshick, S.L.; Chiu, C.-P.; McMurry, P.H. Time-dependent aerosol models and homogenous nucleation rates. *Aerosol Sci. Technol.* **1990**, *13*, 465–477.

7

CHAPTER

Effect of Saturation Pressure Difference on Metal–Silicide Nanopowder Formation in Thermal Plasma Fabrication

Masaya Shigeta [1,*] and Takayuki Watanabe [2]

[1]*Joining and Welding Research Institute, Osaka University, 11-1 Mihogaoka, Ibaraki, Osaka 567-0047, Japan*

[2]*Department of Chemical Engineering, Kyushu University, 744 Motooka, Nishi-ku, Fukuoka 819-0395, Japan*

Abstract

A computational investigation using a unique model and a solution algorithm was conducted, changing only the saturation pressure of one material artificially during nanopowder formation in thermal plasma fabrication, to highlight the effects of the saturation pressure difference between a metal and silicon. The model can not only express any profile of particle size–composition distribution for a metal–silicide nanopowder even with widely ranging sizes from sub-nanometers to a few hundred nanometers, but it can also simulate the entire growth process involving binary homogeneous nucleation, binary heterogeneous co-condensation, and coagulation among nanoparticles with different compositions. Greater differences in saturation pressures cause a greater time lag for co-condensation of two material vapors during the collective growth of the metal–silicide nanopowder. The greater time lag for co-condensation results in a wider range of composition of the mature nanopowder.

Keywords: nanopowder; metal silicide; co-condensation; thermal plasma; modeling

1. Introduction

Thermal plasmas have been used for effectual fabrication of nanopowders composed of nanometer-scale particles [1]. Nanopowders have unique capabilities that differ greatly from those of bulk materials or powders composed of larger particles [2]. Particularly, nanopowders composed of metal–silicide nanoparticles are anticipated to be potentially useful materials for extremely small electronic and mechanical applications such as solar-controlled windows, electromagnetic shielding, and contact materials in microelectronics [3]. However, because those raw materials usually have high melting points or boiling points, high-rate fabrication of those nanopowders is almost impossible using conventional methods such as grinding techniques and liquid-phase preparation. Combustion processes are also unusable because they are accompanied by unfavorable production of contaminants attributable to the oxidation atmosphere and because their flames cannot reach sufficiently high temperatures to vaporize the raw materials. Thermal plasmas offer the distinct benefits of high enthalpy, high chemical reactivity, variable properties, and a high cooling rate, all of which suit high-rate fabrication of metal–silicide nanopowders [4]. Additionally, the temperature and flow fields are controllable using external electromagnetic fields [5,6,7].

Thermal plasma fabrication of metal–silicide nanopowders involves the vaporization of raw materials and the subsequent conversion of the binary material vapors into numerous nanoparticles by virtue of the high enthalpy and high cooling rate of thermal plasma. However, the nanopowder growth is tremendously complicated because the nanopowder grows in a few tens of milliseconds through simultaneous and collective processes of binary homogeneous nucleation, binary heterogeneous co-condensation, and coagulation among nanoparticles with different compositions. Therefore, observing the growth process directly during experimentation is impossible. Only the characteristics of the final products have been evaluated [8,9,10,11,12]. Therefore, the growth mechanism remains poorly understood.

Computational studies based on theoretical modelling can reveal the growth mechanism and can enable prediction of the profile of the nanopowder to be synthesized. However, because of computational resource limitations, molecular dynamics (MD) calculation cannot comprehensively treat the entire growth process from nucleation until a nanopowder completes its growth [13]. In place of MD calculation with a heavy computational load, models based on aerosol dynamics have been used to simulate the process of a collective nanopowder growth comprehensively. Nevertheless, most models are applicable only to unary systems [14,15,16,17,18,19,20,21].

Only a few aerosol-dynamics-based models have been developed for thermal plasma fabrication of nanopowders involving co-condensations of binary material vapors [22,23,24]. Those models adopted several oversimplifications to obtain simple numerical solutions including only mean values. For more accurate and detailed numerical analysis in the growth processes of binary material nanopowders, we developed a unique model and solution algorithm [25,26,27,28]. That model can not only express any profile of particle size–composition distribution (PSCD) of a nanopowder even with widely ranging sizes from sub-nanometers to a few hundred nanometers, but can also simulate the entire formation process involving binary homogeneous nucleation, binary heterogeneous co-condensation, and coagulation among nanoparticles with different compositions. Especially for nanopowder formation of metal–silicides (Mo–Si, Ti–Si, Co–Si) under thermal plasma conditions, the model produced numerical results that agreed with experiment results [26,28].

Those computational results showed that the difference in saturation pressures between a metal and silicon was a crucial factor that determined the time lag of co-condensation and consequently affected the range of the silicon content in the synthesized nanopowder. Actually, an experimental study also reported that the saturation pressure difference affected the nanoparticle composition [11]. The literature emphasized that the difference of the saturation pressure caused that of the nucleation temperature and the larger difference resulted in a larger composition range of metal–silicide nanoparticles. Following this experimental study, the effect was investigated computationally using a simpler model as well [29]. Although the model did not consider binary nucleation and coagulation, it also predicted that systems with large differences of saturation pressures tended to produce metal–silicide nanoparticles with a wide range of compositions due to a time-lag of condensations of two materials.

Even though those studies indicated the importance of the saturation pressure difference between a metal and silicon, the effect remained unclear because the formation processes of nanopowders were also affected by material properties other than the saturation pressure. Data from the actual material properties were used for each material in those computations [26,28,29]. In experiments, it is generally impossible to control only a saturation pressure by changing materials. Other properties are changed as well. Therefore, in this study, numerical experiments are performed by changing only the saturation pressure of one material artificially to highlight the effects of saturation pressure differences on the metal–silicide nanopowder formation using the model which can simulate collective formation through simultaneous processes of binary nucleation, binary co-condensation, and coagulation among nanoparticles with different compositions [26,28].

2. Computational Conditions and Strategy

Figure 1 presents a schematic illustration of metal–silicide nanopowder fabrication using induction thermal plasma (ITP). The precursory raw materials are injected into a plasma where the high-temperature field vaporizes materials completely [22]. Metal and silicon vapors are transported with the flow to the plasma's tail, which exhibits a rapid temperature decrease. Consequently, either or both of the material vapors become supersaturated, which engenders homogeneous nucleation. Because it is a binary system, nuclei composed of the metal atoms and silicon atoms are generated (binary nucleation). Immediately, the binary material vapors co-condense heterogeneously on the nuclei (binary co-condensation). Furthermore, during their growth, the nanoparticles mutually collide and merge into larger nanoparticles (coagulation). The metal–silicide nanopowder growth consists of these three processes that progress collectively and simultaneously. As a consequence, such nanopowders always have varieties of sizes and compositions, as shown in the experiments [11,22,28].

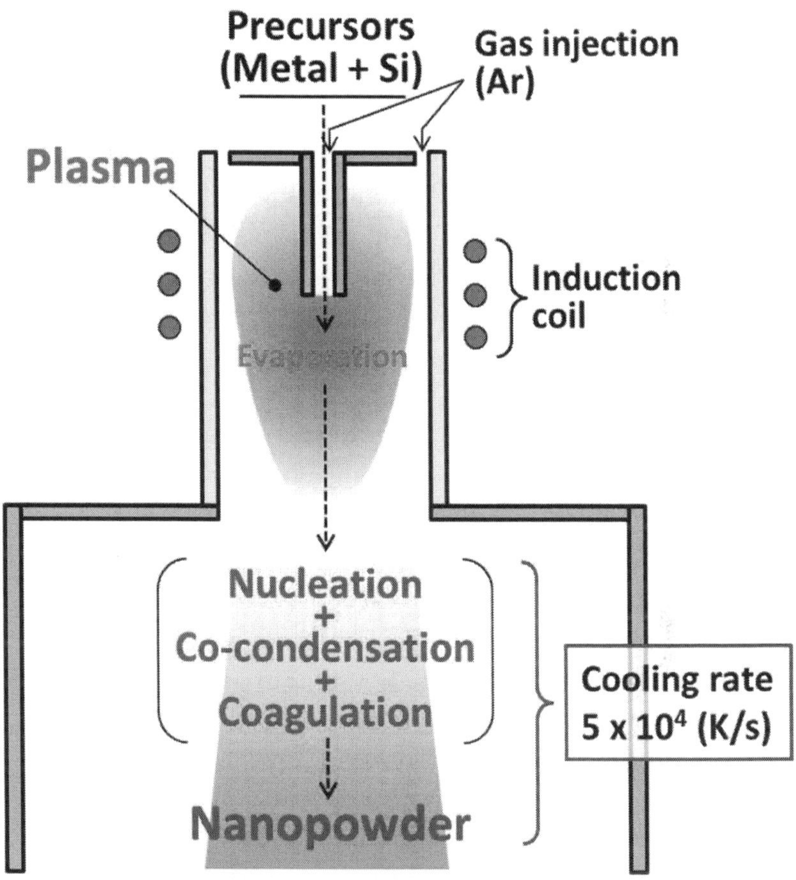

Figure 1. Metal–silicide nanopowder fabrication using an induction thermal plasma.

In a typical condition of ITP discharge, the region downstream from the plasma offers a high cooling rate of 10^4–10^5 K/s. Therefore, the present computation sets a constant cooling rate of 5.0×10^4 K/s to investigate the effect of saturation pressure difference under extremely simple conditions. The initial mole fraction of the material vapor to argon gas is set to be 0.5%. This can be regarded as a dilute condition in which the effect of the raw material on the flow field is negligible. This study particularly selects a titanium–silicon binary system because the saturation pressures of titanium and silicon are mutually close ($p_{S(\mathrm{Ti})}/p_{S(\mathrm{Si})} = 10^{-1}$–$10^0$). The initial ratio of these materials is set fairly at Ti:Si = 1:1. These conditions suggest that the vapors of titanium and silicon co-condense almost simultaneously.

As a great benefit of computational investigation, the value of only one saturation pressure can be changed strategically, which cannot be done in experiments. Therefore, this study defines an artificial saturation pressure $p_S' = \zeta \cdot p_S$. Controlling only the value of ζ from 10^{-3} to 10^3, while using the other actual material properties [30] with no changes, the effect of saturation pressure difference is highlighted. Computations are performed with a time increment Δt of 2.0 μs, which provides sufficient resolution for the present condition.

3. Outline of Metal–Silicide Nanopowder Formation Model

Formation of binary metal-silicide nanopowder from the vapor phase can be computed using a unique model developed by the authors [26,27,28]. The model with the PSCD describes a collective and simultaneous growth process of two-component nanoparticles in a binary vapor system through binary homogeneous nucleation, binary heterogeneous co-

condensation, and coagulation among nanoparticles with different compositions. The PSCD is defined on a 2D coordinate system with two individual variables of the particle size and composition (here, silicon content), where nanoparticles composing a nanopowder are present only at the grid points [26].

In the model, the free energy of cluster formation W [31] is an important variable that dominates the nucleation and co-condensation:

$$W = -n_{(A)}k_BT\ln\left(\frac{N_{mono(A)}}{N'_{S(A)}}\right) - n_{(B)}k_BT\ln\left(\frac{N_{mono(B)}}{N'_{S(B)}}\right) + \sigma' s' \quad (1)$$

where $N_{mono(M)}$ stands for the monomer number density of material M (= A or B), $N'_{S(M)}$ signifies the equilibrium monomer number density of material M in the saturated vapor over a bulk solution. In addition, σ' and s' respectively represent the surface tension and the surface area of the cluster. For binary clusters, σ' is estimated approximately as:

$$\sigma' = \frac{n_{(A)}\sigma_{(A)} + n_{(B)}\sigma_{(B)}}{n_{(A)} + n_{(B)}} \quad (2)$$

Therein, $n_{(M)}$ is monomers of material M contained in a binary cluster and $\sigma_{(M)}$ is surface tension of material M. It is noteworthy that the nanoparticles are allowed to grow by condensation only when the free energy gradients for particle formation, W, is negative or zero. Therefore:

$$\frac{\partial W}{\partial n_{(M)}} \leqslant 0 \quad (3)$$

During a nanopowder growth process with a temperature decrease, the nanoparticles will be solidified. Then they can no longer increase their size as spherical particles by coagulation. In general, the solidification point depends on the material composition [32]. Furthermore, the solidification point decreases with the particle diameter [33]. Although these effects of the solidification point variation of binary-component nanoparticles were considered in our previous studies [26,27,28], the present study removes these effects to clarify only the effect of saturation pressure difference on the growth process.

4. Results and Discussion

Figure 2 shows the PSCD evolution of the Ti–Si nanopowder for $\zeta = 1$, which describes the growth process in an actual Ti–Si binary system. Figure 3a presents the histories of the vapor pressures and the saturation pressures of Ti and Si, whereas Figure 3b depicts the conversion ratios that indicate how much of each material vapor has been converted into nanoparticles. It is noted that the horizontal axes show the temperature in the opposite direction because of the cooling process. According to an earlier study [34], the nucleation rate with 1 nucleus/cm^3s can be conveniently observed experimentally. In this study, the nucleation is judged to start when the nucleation rate first exceeds this value. For the present case, nucleation starts when the vapors of Ti and Si are cooled to a temperature lower than 2444 K.Figure 2a shows that nuclei composed of Ti and Si are generated at the early stage of growth. At this time, the vapors of both Ti and Si are supersaturated, as shown in Figure 3a. Following nucleation, the nanoparticles grow rapidly (Figure 2b–e) by the simultaneous co-condensation of the material vapors on the nuclei as portrayed in Figure 3b as well as coagulation among themselves. Figure 3b also shows that 99% of Ti vapor completes the conversion at 2055 K, whereas 99% of Si vapor completes the conversion at 1977 K. After this drastic growth, the nanopowder grows slowly through coagulation and finally reaches its mature state of Figure 2f at 1716 K. To determine such a mature state of a nanopowder, the maximum difference of the particle number density at each node for Δt was monitored. That monitored parameter was defined as:

$$Q = \max\left(\frac{\left|\hat{N}_{i,j}^{(t)} - \hat{N}_{i,j}^{(t-\Delta t)}\right|}{\hat{N}_{i,j}^{(t-\Delta t)}}\right) \tag{4}$$

where

$$\hat{N}_{i,j}^{(t)} = \frac{N_{i,j}^{(t)}}{\rho_g^{(t)}} \tag{5}$$

and $\rho_g^{(t)}$ is the bulk gas density at the time t. When Q fell to less than 0.1, the nanopowder was determined to be mature. The mature nanopowder is composed mainly of the nanoparticles with the silicon content of $x_{(Si)}$ = 50.0 atom %, which is identical to the initially given silicon fraction to the precursor. Particle diameters range widely from a few nanometers to 76 nm.

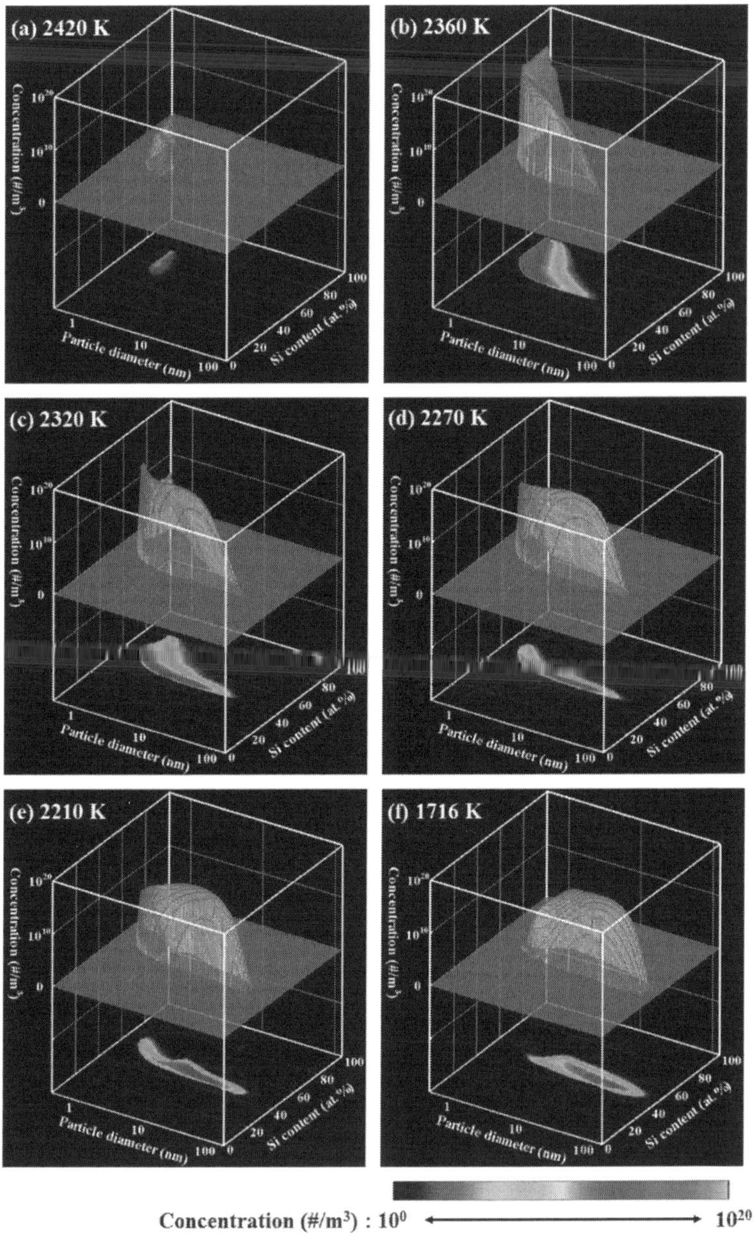

Figure 2. PSCD evolution for $\zeta = 1$.

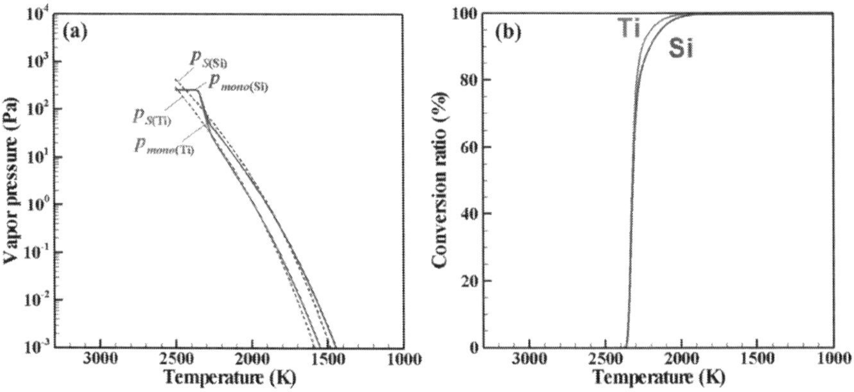

Figure 3. Phase conversion histories for ζ = 1: (**a**) vapor pressures and (**b**) conversion ratios.

Figure 4 shows the PSCD evolution of the Ti–Si nanopowder for ζ = 10^3. This value means that the saturation pressure of silicon is artificially set to be 1,000 times larger than the actual saturation pressure. Figure 5a,b respectively indicate the histories of the vapor pressures and the conversion ratios. Nucleation starts at 2258 K. At the early stage of growth, Ti-rich nuclei are generated (Figure 4a) and Ti-rich nanoparticles are formed (Figure 4b) only by Ti vapor condensation because the Si vapor pressure is still much lower than the saturation pressure (Figure 5a). During this growth of the Ti-rich nanoparticles, Si vapor starts to co-condense on the Ti-rich nanoparticles with Ti vapor (Figure 4c–e).Figure 5a shows that 99% of Ti vapor completes the conversion at 1882 K. After this co-condensation, Si vapor continues to condense on the Ti–Si nanoparticles slowly; 99% of Si vapor is consumed at 1506 K. The nanopowder reaches its mature state shown in Figure 4f at 1408 K. Although most the mature nanopowder exhibits silicon contents of $x_{(Si)}$ = 50.0 atom %, the composition ranges widely from 25 atom % to 80 atom %.

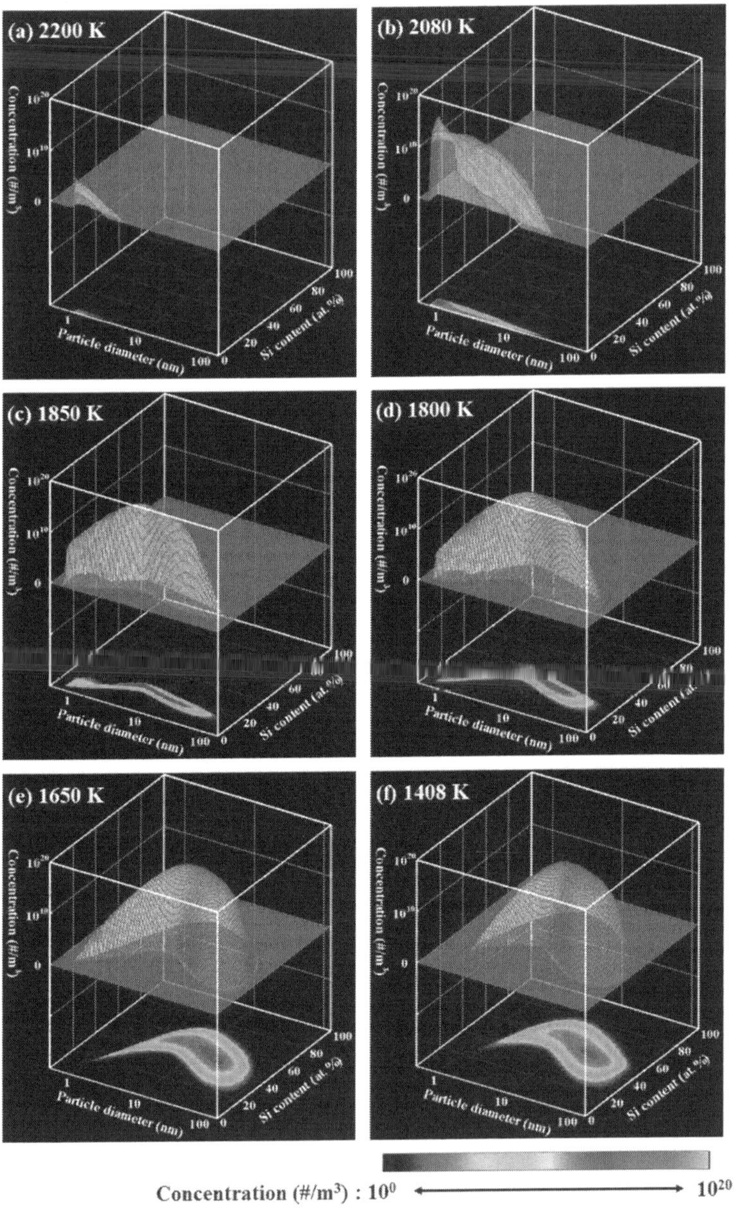

Figure 4. PSCD evolution for $\zeta = 10^3$.

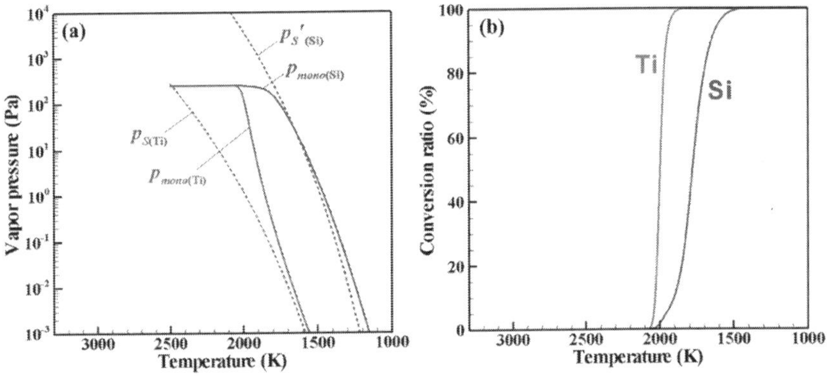

Figure 5. Phase conversion histories for ζ = 10^3: (**a**) vapor pressures and (**b**) conversion ratios.

Figure 6 shows the PSCD evolution of the Ti–Si nanopowder for ζ = 10^{-3}. Figure 7a,b respectively depict the vapor pressures and the conversion ratios. Nucleation starts at 3777 K, which is much higher than in the other cases. At the early stage of the growth, Si-rich nuclei are generated (Figure 6a) and Si-rich nanoparticles are formed (Figure 6b) by Si vapor condensation because only the Si vapor pressure is supersaturated (Figure 7a). Following this growth of Si-rich nanoparticles, Ti vapor starts to co-condense on the Si-rich nanoparticles with Ti vapor (Figure 6c–e). 99% of Si vapor is consumed at 2654 K, at which only 9% of Ti vapor is consumed (Figure 7b). 99% of Ti vapor completes the conversion at 2031 K. The nanopowder reaches its mature state as seen in Figure 6f at 1894 K. This mature nanopowder also has widely ranging silicon contents, from 25 atom % to 80 atom %.

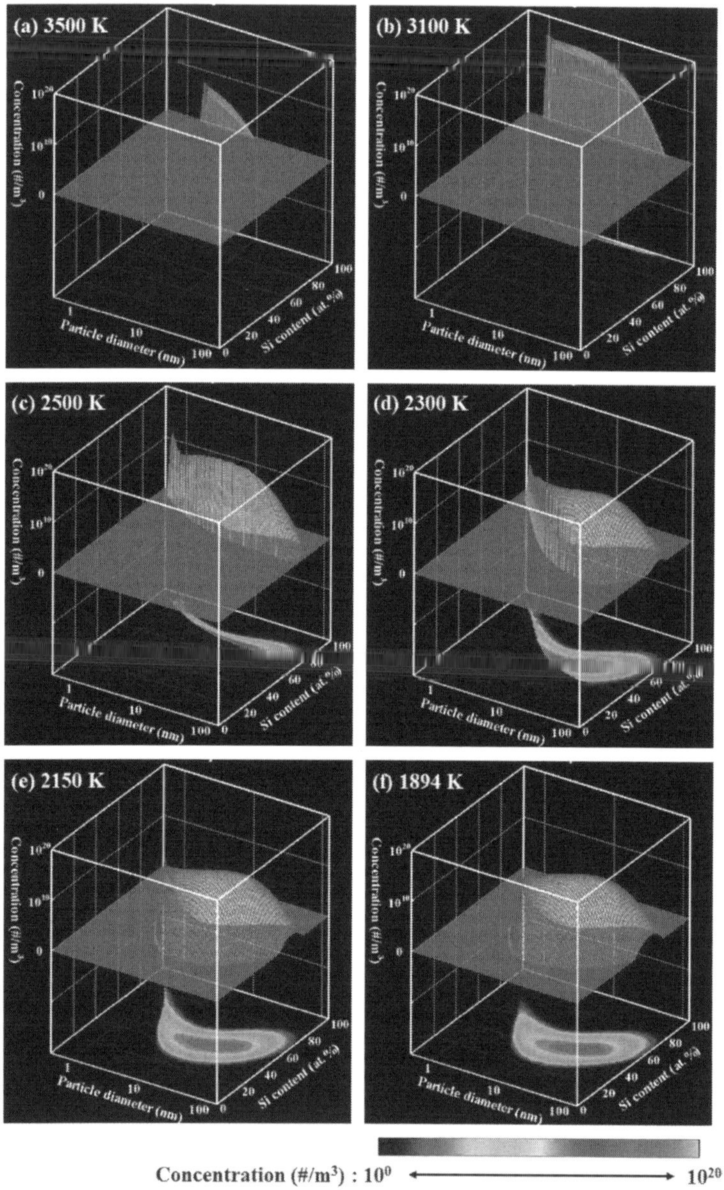

Figure 6. PSCD evolution for $\zeta = 10^{-3}$.

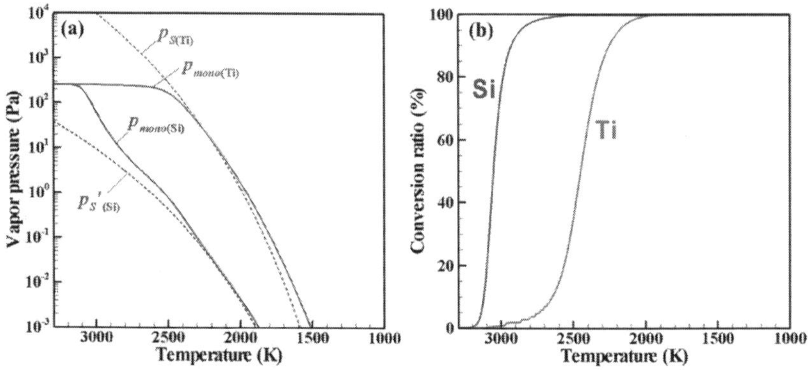

Figure 7. Phase conversion histories for ζ = 10^{-3}: (**a**) vapor pressures and (**b**) conversion ratios.

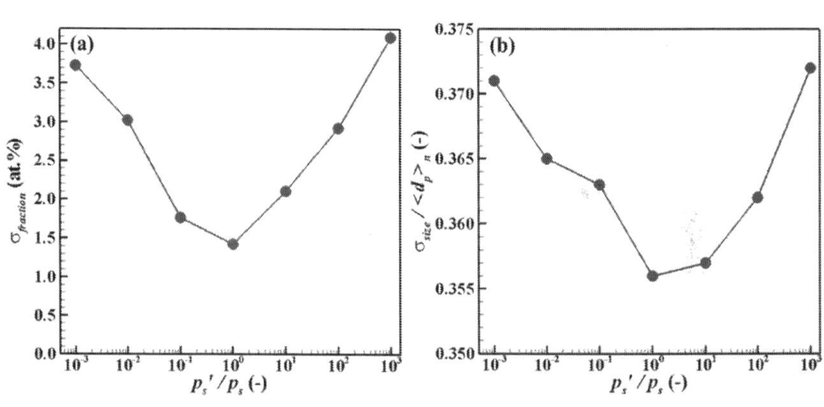

Figure 8. Effects of saturation pressure difference on dispersion of (**a**) fraction and (**b**) size.

These results imply that a wider range of composition is caused by a larger time lag of co-condensation that originally results from a larger difference of saturation pressures. Figure 8 shows the results obtained in the same manner using numerical experiments. As presumed, the standard deviations of fraction are larger when the saturation pressure differences are larger. Additionally, the results show that the standard deviations of size normalized by the arithmetic mean diameters are larger when the saturation pressure differences are larger.

5. Conclusions

Using our unique model and solution algorithm, which can simulate the entire growth process involving binary homogeneous nucleation, binary heterogeneous co-condensation, and coagulation among nanoparticles with different compositions, numerical experiments were carried out by changing only the saturation pressure of one material artificially to highlight the effect of the saturation pressure difference between a metal and silicon on the nanopowder formation in thermal plasma fabrication. A greater difference in saturation pressures causes a larger time lag of co-condensation of two material vapors during the collective growth of the metal–silicide nanopowder. The longer time lag of co-condensation results in a wider range of compositions for the mature nanopowder.

Acknowledgments

This work was partly supported by the Japan Society for the Promotion of Science Grant-in-Aid for Scientific Research (B) (KAKENHI: Grant No. 15H03919).

Author Contributions

Masaya Shigeta and Takayuki Watanabe conceived and designed the computational experiment. Masaya Shigeta developed the model and algorithm and performed the computation. Masaya Shigeta and Takayuki Watanabe analyzed the data. Masaya Shigeta wrote the paper.

Conflicts of Interest

The authors declare no conflict of interest.

References

1. Shigeta, M.; Murphy, A.B. Thermal Plasmas for Nanofabrication. *J. Phys. D Appl. Phys.* **2011**, *44*, 174025.

2. Siegel, R.W. Synthesis and properties of nanophase materials. *Mater. Sci. Eng.* **1993**, *A168*, 189–197.

3. Matsuura, K.; Hasegawa, T.; Ohmi, T.; Kudoh, M. Synthesis of $MoSi_2$-$TiSi_2$ Pseudobinary Alloys by Reactive Sintering. *Metall. Mater. Transact. A* **2000**, *31*, 747–753.

4. Kambara, M.; Kitayama, A.; Homma, K.; Hideshima, T.; Kaga, M.; Sheem, K.-Y.; Ishida, S.; Yoshida, T. Nano-composite Si particle formation by plasma spraying for negative electrode of Li ion batteries. *J. Appl. Phys.* **2014**, *115*, 143302.

5. Sato, T.; Shigeta, M.; Kato, D.; Nishiyama, H. Mixing and magnetic effects on a nonequilibrium argon plasma jet. *Int. J. Therm. Sci.* **2001**, *40*, 273–278.

6. Shigeta, M.; Sato, T.; Nishiyama, H. Computational simulation of a particle-laden RF inductively coupled plasma with seeded potassium. *Int. J. Heat Mass Transf.* **2004**, *47*, 707–716.

7. Shigeta, M.; Nishiyama, H. Numerical Analysis of Metallic Nanoparticle Synthesis Using RF Inductively Coupled Plasma Flows. *J. Heat Transf.* **2005**, *127*, 1222–1230.

8. Fan, X.; Ishigaki, T. Critical free energy for nucleation from the congruent melt of $MoSi_2$. *J. Cryst. Growth* **1997**, *171*, 166–173.

9. Fan, X.; Ishigaki, T.; Sato, Y. Phase formation in molybdenum disilicide powders during in-flight induction plasma treatment. *J. Mater. Res.* **1997**, *12*, 1315–1326.

10. Watanabe, T.; Itoh, H.; Ishii, Y. Preparation of ultrafine particles of silicon base intermetallic compound by arc plasma method. *Thin Solid Films* **2001**, *390*, 44–50.

11. Watanabe, T.; Okumiya, H. Formation mechanism of silicide nanoparticles by induction thermal plasmas. *Sci. Technol. Adv. Mater.* **2004**, *5*, 639–646.

12. Gerile, N.; Kaga, M.; Kambara, M. Synthesis and Characterization of the Plasma Sprayed Si–Ni Composite Powders as Negative Electrode of lithium-ion Batteries. In Proceedings of the 12th Asia Pacific Physics Conference (APPC12), Makuhari, Japan, 14–19 July 2013.

13. Lümmen, N.; Kraska, T. Homogeneous nucleation and growth in iron-platinum vapor investigated by molecular dynamics simulation. *Eur. Phys. J. D* **2007**, *40*, 247–260.

14. Girshick, S.L.; Chiu, C.-P.; Muno, R.; Wu, C.Y.; Yang, L.; Singh, S.K.; McMurry, P.H. Thermal plasma synthesis of ultrafine iron particles. *J. Aerosol Sci.* **1993**, *24*, 367–382.

15. Bilodeau, J.F.; Proulx, P. A mathematical model for ultrafine iron powder growth in thermal plasma. *Aerosol Sci. Technol.* **1996**, *24*, 175–189.

16. Desilets, M.; Bilodeau, J.F.; Proulx, P. Modelling of the reactive synthesis of ultra-fine powders in a thermal plasma reactor. *J. Phys. D Appl. Phys.* **1997**, *30*, 1951–1960.

17. Cruz, A.C.D.; Munz, R.J. Vapor Phase Synthesis of Fine Particles. *IEEE Transact. Plasma Sci.* **1997**,*25*, 1008–1016.

18. Murphy, A.B. Formation of titanium nanoparticles from a titanium tetrachloride plasma. *J. Phys. D Appl. Phys.* **2004**, *37*, 2841–2847.

19. Shigeta, M.; Watanabe, T. Two-dimensional analysis of nanoparticle formation in induction thermal plasmas with counterflow cooling. *Thin Solid Films* **2008**, *516*, 4415–4422.

20. Shigeta, M.; Watanabe, T. Numerical investigation of cooling effect on platinum nanoparticle formation in inductively coupled thermal plasmas. *J. Appl. Phys.* **2008**, *103*, 074903.

21. Colombo, V.; Ghedini, E.; Gherardi, M.; Sanibondi, P.; Shigeta, M. A Two-Dimensional nodal model with turbulent effects for the synthesis of Si nano-particles by

inductively coupled thermal plasmas. *Plasma Sources Sci. Technol.* **2012**, *21*, 025001.

22. Shigeta, M.; Watanabe, T. Growth mechanism of silicon-based functional nanoparticles fabricated by inductively coupled thermal plasmas. *J. Phys. D Appl. Phys.* **2007**, *40*, 2407–2419.

23. Vorobev, A.; Zikanov, O.; Mohanty, P. Modelling of the in-flight synthesis of TaC nanoparticles from liquid precursor in thermal plasma jet. *J. Phys. D Appl. Phys.* **2008**, *41*, 085302.

24. Vorobev, A.; Zikanov, O.; Mohanty, P. A Co-Condensation Model for In-Flight Synthesis of Metal-Carbide Nanoparticles in Thermal Plasma Jet. *J. Therm. Spray Technol.* **2008**, *17*, 956–965.

25. Shigeta, M.; Watanabe, T. Two-Directional Nodal Model for Co-Condensation Growth of Multi-Component Nanoparticles in Thermal Plasma Processing. *J. Therm. Spray Technol.* **2009**, *18*, 1022–1037.

26. Shigeta, M.; Watanabe, T. Growth model of binary alloy nanopowders for thermal plasma synthesis. *J. Appl. Phys.* **2010**, *108*, 043306.

27. Cheng, Y.; Shigeta, M.; Choi, S.; Watanabe, T. Formation Mechanism of Titanium Boride Nanoparticles by RF Induction Thermal Plasma. *Chem. Eng. J.* **2012**, *183*, 483–491.

28. Shigeta, M.; Watanabe, T. Effect of precursor fraction on silicide nanopowder growth under thermal plasma

conditions: a computational study. *Powder Technol.* **2016**, *288*, 191–201.

29. Shigeta, M.; Watanabe, T. Numerical analysis for co-condensation processes in silicide nanoparticle synthesis using induction thermal plasma at atmospheric pressure conditions. *J. Mater. Res.* **2005**, *20*, 2801–2811.

30. The Japan Institute of Metals. *Metal Data Book*; Maruzen: Tokyo, Japan, 1993; pp. 10–91. (In Japanese)

31. Wyslouzil, B.E.; Wilemski, G. Binary nucleation kinetics. II. Numerical solution of the birth–death equations. *J. Chem. Phys.* **1995**, *103*, 1137–1151.

32. Massalski, T.B. *Binary Alloy Phase Diagrams*, 2nd ed.; American Society for Metals: Materials Park, OH, USA, 1990; pp. 3367–3371.

33. Wautelet, M.; Dauchot, J.P.; Hecq, M. Phase diagrams of small particles of binary systems: A theoretical approach. *Nanotechnology* **2000**, *11*, 6–9.

34. Friedlander, S.K. *Smoke, Dust and Haze, Fundamentals of Aerosol Dynamics*, 2nd ed.; Oxford University Press: New York, NY, USA, 2000; p. 280.

8

CHAPTER

Restorative Self-Image of Rough-Line Grids: Application to Coherent EUV Talbot Lithography

Hyun-su Kim

Chair for the Experimental Physics of EUV, RWTH Aachen University and Jülich Aachen Research Alliance, Fundamentals for Future Information Technology (JARA-FIT), 52074 Aachen, Germany

ABSTRACT

Self-imaging is a well-known optical phenomenon produced by diffraction of a coherent beam in a periodic structure. The self-imaging effect (or Talbot effect) replicates the field intensity at a periodic mask in certain planes, effectively producing in those planes an image of the mask. However, the effect has not been analyzed for a rough-line grid from the point of view of the fidelity of the image. In this paper, we investigate the restorative effect of the self-image applied to the lithography of gratings with rough lines. This paper is applied to characterize a Talbot lithography experiment implemented in the extreme ultraviolet. With the self-imaging technique, a mask with grid patterns having bumps randomly placed along the line edges reproduces a grid pattern with smoothed line edges. Simulation explores the approach further for the cases of sub-100-nm pitch grids.

Keywords: Talbot effect, nanofabrication, Extreme ultraviolet application

INTRODUCTION

In this work, we analyze the restoration effect produced in Talbot images applied to nanolithography. Talbot lithography in the extreme ultraviolet (EUV) has been demonstrated as a convenient method to fabricate nanostructures. The technique is of a great interest for many applications, requiring well-defined periodic structures, such as defect free solar cells, semiconductor photonic devices fabrication, or pre-patterning for self-assembled growth of quantum dots (QDs) arrays [1].

The Talbot effect, or self-imaging, is a well-known optical phenomenon. When a coherent plane wave impinges on a periodic object such as a grating, an image is generated at regular distances away from the object. This distance, i.e., the so-called Talbot distance (z_T), is determined by

$$z_T = 2 \cdot d^2/\lambda \quad (1)$$

where is the pitch (period) of the object structures, and is the wavelength of illumination [2], [3]. The Talbot pattern is repeated in the planes located at distances $z = n \cdot z_T$, where n is a positive integer. The self-imaging effect can be utilized in lithography to replicate nanoscale periodic structures illuminating a periodic mask with coherent EUV radiation. This approach called EUV Talbot lithography has been examined in many aspects, e.g., showing its capability to replicate complex patterns [4], the possibility to obtain spatial frequency multiplication [5], and the characteristic of defect tolerant printing [6].

The quality of the patterned structures is important, because it influences the functionality of the device. The fidelity of the printed pattern is particularly critical in the fabrication of micro-electronics components. The microelectronic industry had invested significant amount of effort and resources to mitigate this problem, localizing and repairing defects in the lithography masks. However, in addition, in other applications besides the main stream of the micro-components industry, the quality of the lithography print is instrumental to a successful device fabrication and performance.

Talbot lithography is particularly well adapted to produce high quality lithography prints with high resolution over large areas. It has been shown that the Talbot effect can reproduce the defect free images from a defective periodic mask. This filtering characteristic was identified as a convenient way to improve the quality of the print in a lithography process [7], [8]. The defect tolerance characteristic of the Talbot images is well studied in the bibliography. It was shown that a mask with a periodic pattern, which is locally damaged, produces a defect-free copy at the Talbot distance. The effect of a defect on the pattern size around the defect in ArF Talbot lithography was investigated for submicron patterns [9]. In this case, the periodic mask included regularly distributed defects. On the other hand, it was shown how the presence of spherical particles behind gratings affects the formation of Talbot self-images [10]. The tolerance in angles of continuously self-imaging gratings was studied for nonparaxial illumination both numerically and experimentally [11]. The defect tolerant Talbot imaging was studied for lithography applications with a EUV laser in the case of masks with complex designs [6], [12]. For the particular case of a line grating, it is increasingly difficult to fabricate a straight-line edge when the pitch is reduced. Thus, it would be worth to analyze the restoration effect of the Talbot image in this case, quantifying the quality of the replicated pattern for the rough-line grid. In this work, we analyze with numerical simulations and experimental results the effectiveness of EUV Talbot lithography to improve the lithography print from rough line grid masks. For the test we used a binary transmission mask fabricated in an opaque membrane with slits opened through it and a coherent EUV high-photon flux laser. We observed improvement in the pattern recorded in the photoresist compared with the original pattern on the mask. We also performed a model calculation of the obtained image for sub-100 nm pitches and analyzed the limits of the improvement.

SIMULATED DEMONSTRATION

Fig. 1 illustrates the scheme of the self-imaging effect. When the mask [Fig. 1(a)] is illuminated by a plane wave, a replica of the mask's pattern is observed at the Talbot plane z_T [Fig. 1(b)]. For the simulation, a line grid pattern with rough edges was created. Bumps of ~100 nm radius were randomly distributed along the line edges. The size of the bumps along the edges of the slits that produce the roughness was selected to be bigger than the cut-off size defined as $\Delta x \approx \lambda/2\pi \approx 7.5$ nm for the wavelength at 46.9 nm. This selection validates the results of the scalar theory used in the simulations. The distribution was performed using the random function in MATLAB's built-in functions. The diffraction pattern was calculated using the Fresnel diffraction formalism, by the fast Fourier transform method [13]. The scalar diffraction method assumes the mask as the thin mask approximation (TMA). The simulation is performed in a finite grid of $\Delta x = 0.5$ nm and $\Delta y = 0.5$ nm over a field of $50\ \mu m \times 50\ \mu m$ in the mask. Fig. 2 shows the simulated images of (a) a mask with randomly rough lines and (b) the self-image at z_T.

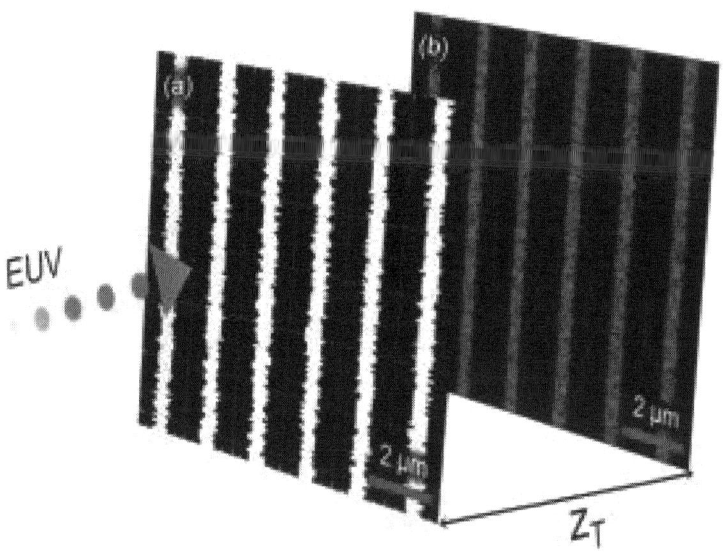

Fig. 1. Scheme of the self-imaging setup. (a) Talbot mask. (b) Self-image produced at a distance .

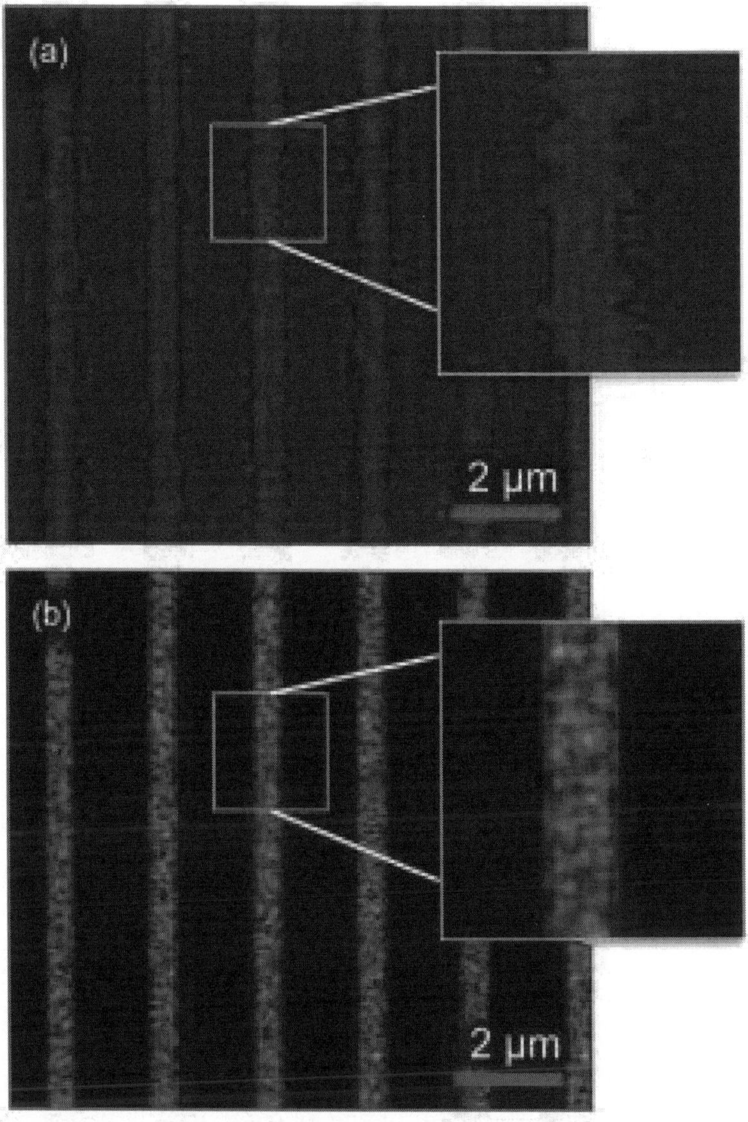

Fig. 2. Simulated images of (a) a mask with randomly rough lines and (b) the self-image at z_T.

The simulation shows that the self-image at z_T has significantly better quality, in which the bumps along edges are smoothed out over the grid. As a consequence the rough lines were straightened in the self-image. The roughness along an edge was quantified using standard deviation (σ) defined as

$$\sigma = \sqrt{\frac{1}{N} \cdot \sum_{i=1}^{N} (x_i - \overline{x})^2} \qquad (2)$$

where *N* is the number of samplings along a line, and the quantity in the bracket is the deviation of the edge of the slit from the straight line at the sampling point i. In the case of the self-image, the deviation of the edge is determined after a binary thresholding of the calculated image. Consequently the deviation of the edge depends on the value selected for the threshold. Fig. 3 shows the calculated for different threshold values ranging from 0.05 to 0.7. The calculated $\sigma = 95$ nm on the mask [see Fig. 2(a)] was reduced to $\sigma \cong 32$ nm in the Talbot pattern [Fig. 2(b)]. In the analysis, the parabolic curve of in the Talbot pattern occurred due to the irregular intensity distribution of the areal image of Talbot pattern. When the threshold value is either too high or low, the difference of intensity value between neighboring can be very large. This causes the higher σ values in the areal image as compared with the value obtained for the mask.

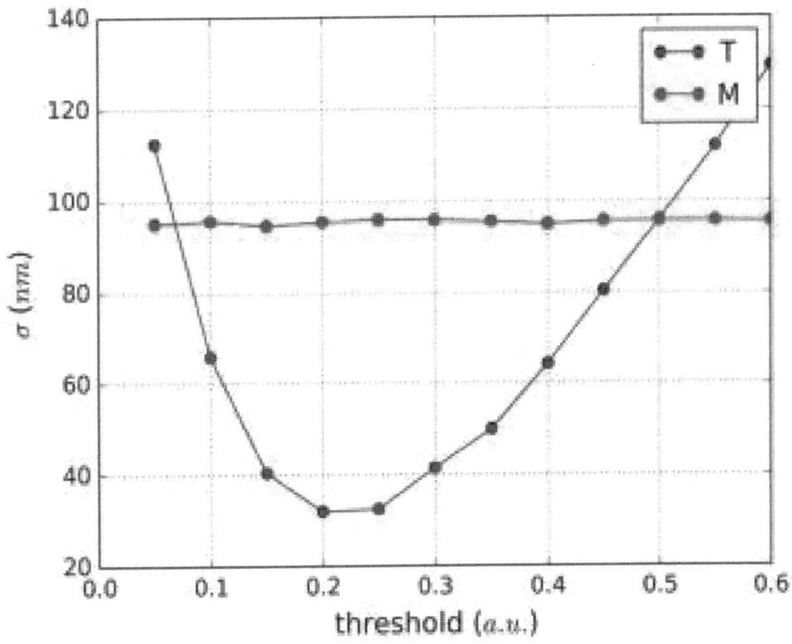

Fig. 3. The values on the self-image (T) in dependence of the threshold level and on the mask $(\breve{M})$ in simulation.

EXPERIMENTAL DEMONSTRATION

The experimental demonstration was performed with an EUV laser and a transmission Talbot mask with a setup as following. The transmission mask and the wafer are aligned in the optical axis as illustrated in Fig. 1. The monochromatic EUV illumination of wavelength 46.9 nm was generated by a capillary discharge plasma laser [14]. The capillary discharge plasma produces a highly spatial and temporal coherent beam well suited for the application described in this work. The spatial coherence length at the position where the self image was obtained was around 350 , which covered the whole area of the mask, assuring a fully spatial coherent illumination. The temporal coherence of the laser is $\Delta\lambda/\lambda \approx 3.5 \times 10^{-5}$ [15]–[17]. [15] [16] [17]. The exposure dose at the sample plane was about 0.1 0.3 mJ/cm^2 per pulse [12], [18].

The transmission Talbot mask was defined using a focused ion beam (FIB) tool to drill through a low stressed Si_3N_4 membrane. The same binary file that was used for the simulation was fed into the FIB to make a more precise comparison between the calculation and the experimental results. After the milling, the mask was deposited with a layer of 50 nm of gold to improve the absorption in the EUV. The pitch in the mask was 2 μm, and the slit width was ~500 nm in average. The illumination field area of the mask was 50 μm × 50 μm.. Fig. 4 shows the SEM image of the fabricated mask.

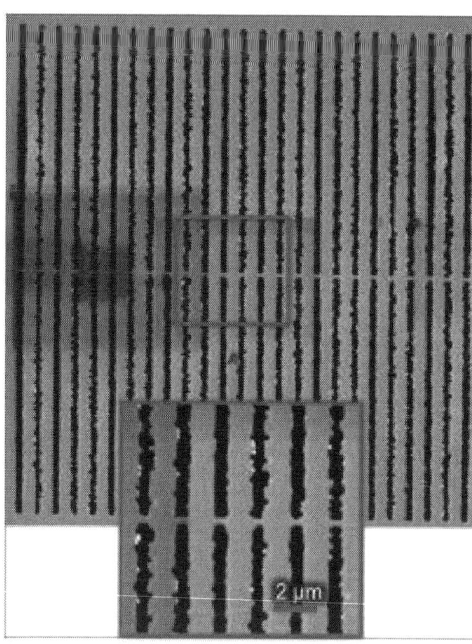

Fig. 4. Free-standing transmission mask having the rough lines grid.

The depth of focus (DOF) in the Talbot plane is expressed as

$$DOF = \frac{\lambda}{(NA)^2} = \lambda\left(1 + 4n^2\left(\frac{d^2}{\lambda \cdot A}\right)^2\right) \quad (3)$$

where the is the area of the mask [4]. With the experimental parameters used the DOF is ~600 nm for the first Talbot plane, $n = 1$.

The print was recorded on a silicon wafer spin-coated with 50 nm thick layer of photoresist (JSR Inc.). During the exposure, the distance from the wafer to mask was maintained close to the first Talbot distance. After the exposure, the exposed photoresist was developed for 30 seconds and rinsed with 2-Propanol. Fig. 5 shows SEM images of the patterned photoresist. The quality of the grid pattern is improved when the wafer is close to (a), (c), rather than in the case when the wafer is out of focus (b), (d). The quality improvement is significant when the distance is precisely controlled to be the Talbot distance. The quality also depends on the exposure time, because the threshold of intensity impacts the development results, as shown in Fig. 5(a), (c) and (b), (d) that the photoresists were exposed with 75 and 110 shots, respectively. Fig. 6(a) is the zoomed-in image of Fig. 5(c). Fig. 6(b) is an AFM image taken at the central part of sample. The roughness is significantly decreased in the print as compared to the roughness in the mask.

Fig. 5. Patterned photoresist images by self-imaging technique. The exposure dose and the distance cause the different qualities in the result.

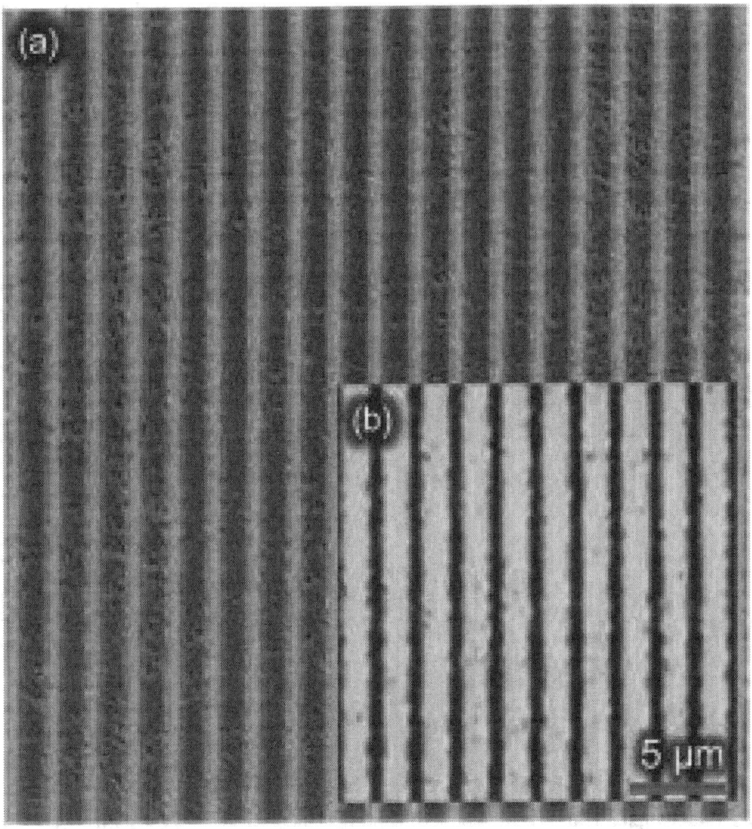

Fig. 6. Patterned photoresist measured with SEM and AFM (right bottom image).

RESULTS AND DISCUSSIONS

In order to compare the roughness in the mask and the one in the self-image, the images shown in Figs. 4 and 6(b) were converted into binary images as shown in Fig. 7. In this case, the gray-scale image turns into the binary image by the balanced histogram thresholding method (BHT). The pixels fall within a desired range

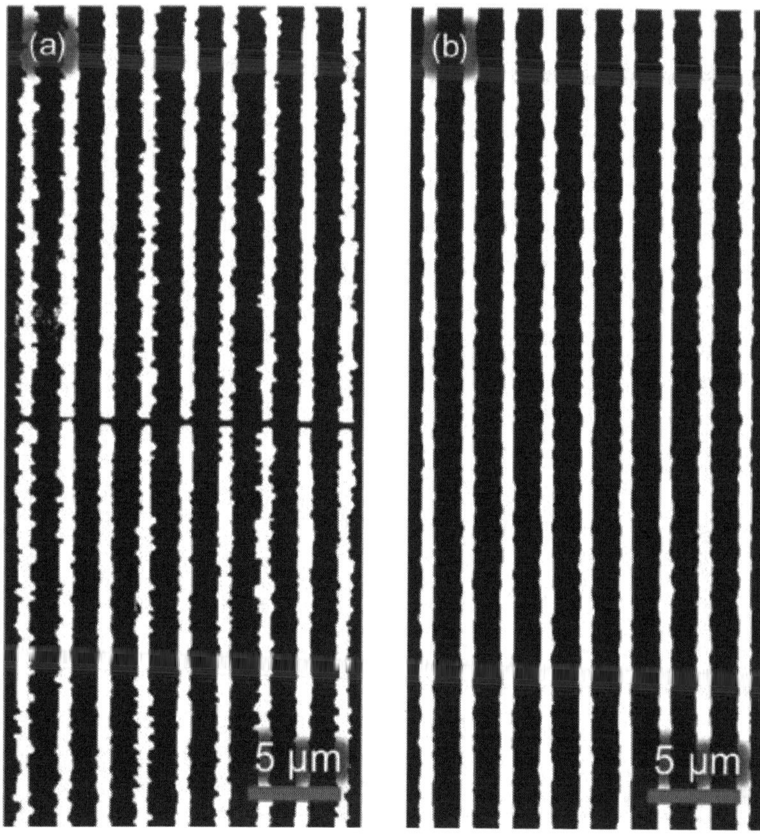

Fig. 7. (a) Mask image and (b) self-image after the binary image conversion.

of intensity value, thus the image is divided to black or white from two main classes: background and foreground [19]. The improvement of quality can be seen in Fig. 7(a) and (b). This apparent improvement correlates well with the reduction of the . Fig. 8 plots the calculated for the mask and the Talbot image for 18 locations along the edges of the slits. The measured in the print are significantly lower than those in mask. The average values are $\sigma = 88$ nm in mask and $\sigma = 2$ nm in the print.

The σ measured in the experimental result correlates well with the obtained in the simulation assuming a threshold for the binarization in the range 0.15–0.3 in Fig. 3. In both simulation and experiment, the values are reduced by about a factor of 2 or more.

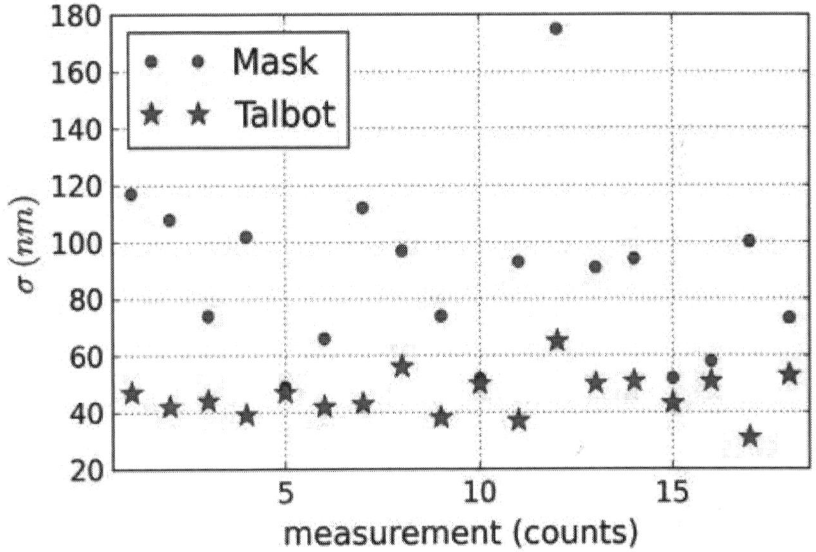

Fig. 8. The σ values in the mask and Talbot patterns.

We also analyzed the influence of the pitch, the slit width and the of the mask to the quality of the print. The study has completed by measuring the roughness of the Talbot image obtained with different masks.

First, to evaluate the influence of the pitch, we tested three different masks. The mask pitches were $p = 100$ nm, $p = 60$ nm, $p = 30$ nm. In those pitches, the roughness was $\sigma = 6.2$ nm, $\sigma = 3.9$ nm, and $\sigma = 2$ nm respectively. Due to the averaging characteristic of the Talbot effect, the σ in the self-image became $\sigma > 3$ nm, $\sigma > 2$ nm, and $\sigma > 1$ nm respectively. The σ in the mask (*M*) and at the Talbot plane (*T*) for selected pitches are plotted in Fig. 9.

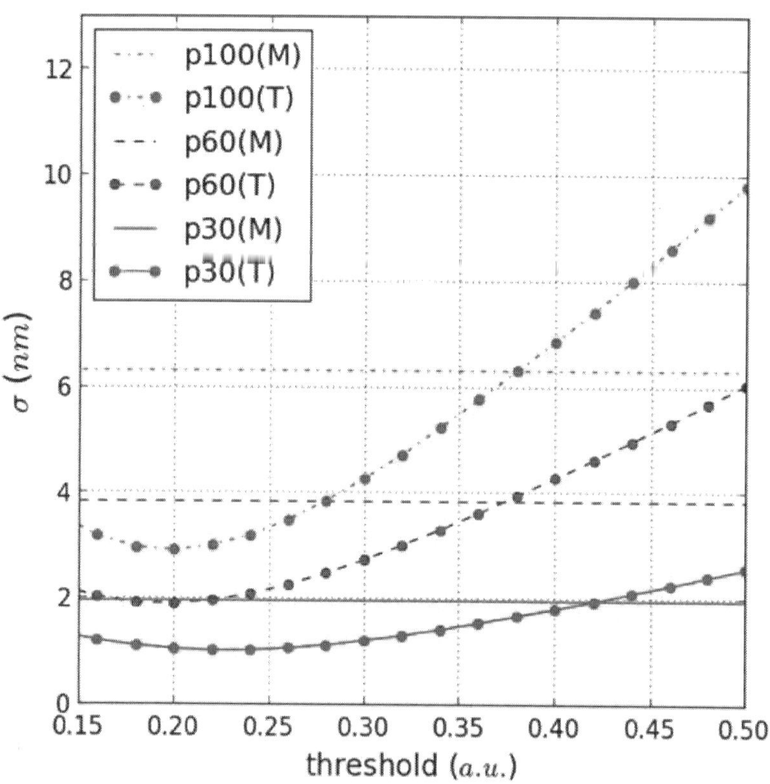

Fig. 9. Simulation result: σ curves in masks (*M*) and in self-images (*T*) for the pitches of 100, 60, and 30 nm.

The complete lithography method includes processes that determine the quality of the print. The printed result is influenced by the photoresist response and the development process among other factors. However, the analysis presented here with the thresholding method in the areal image is intended to study only the self-healing effect in the optical image. As discussed in Fig. 3, when we assume a threshold around 0.3 the simulation and the experimental results are in good agreement. Using this value for the threshold, the parameter in the self-image becomes 4 nm, 2.1 nm and 1.1 nm, respectively. This represents a significant improvement of the self-imaging for sub-100 nm pitch gratings. The roughness is reduced by ~50% for all three masks with different pitches and initial values of the sigma. This amount of improvement is expected for all pitches considered in this analysis.

Second, we tested various mask pitches with identical σ. From this simulation, we used EUV illumination at 13.5 nm wavelength, which is a candidate for the source of the next generation of optical lithography. The pitches selected for this test were 100 nm, 60 nm and 40 nm, with in the three cases fixed at 3.2 nm in mask. As shown in Fig. 10(a), the in the self-image is decreased from ~2 nm to ~1 nm as the pitch is increased from 40 nm to 100 nm. Our analysis indicates that the resulting sigma in the Talbot image is proportional to the ratio of the initial sigma and the slit-width (pitch/2) in the mask. Therefore, in Fig. 10(a), the sigma for the mask with high-dense grid is increased in comparison to the one for the mask with low-dense grid.

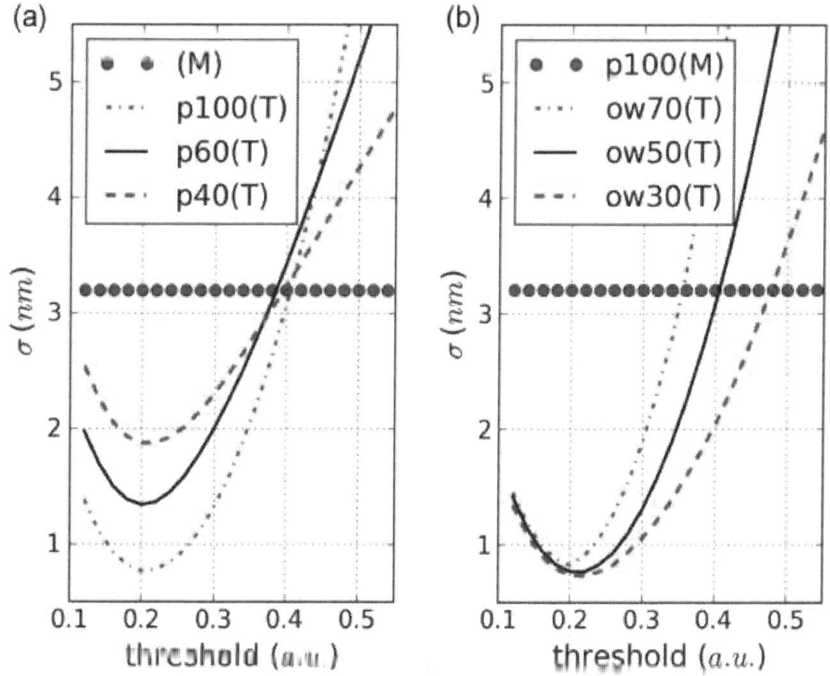

Fig. 10. Simulation results for various parameters. (a) The σ at z_T (T) for pitches of 100, 60, and 40 nm when an identical (M) in mask is used. (b) The σ at z_T (T) for opening widths of 100, 70, and 50 nm when a fixed 100-nm pitch and $\sigma = 3.2$ nm in the mask are used.

Lastly, the slit-width was varied with fixed pitch and in the mask. The values of the slit-width were 70 nm, 50 nm, and 30 nm, where a fixed pitch of 100 nm and a fixed $\sigma = 3.2$ nm in the mask. As can be seen in Fig. 10(b), the minimum values in the self-image were ~0.9 nm. However the dependence on thresholding is different in the three cases enabling wider threshold values for smaller slit widths. For example, for $\sigma = 2$ nm, the threshold increases as the slit width (ow) is decreased as shown in Fig. 10(b). This implies that the fabrication requirements can be relaxed with adjustment of the slit-width. Those results can be of a great interest in nano-patterning, where the line-edge roughness (LER) is one of the significant issues in electronic device fabrication.

CONCLUSION

In conclusion, we have demonstrated a submicron printing technique based on self-imaging and performed a detailed characterization of the quality improvement in the print relative to the mask. The rough grid pattern on the mask produces a self-image with line-edge roughness reduced by about a factor of 2 for 2 to 30 nm pitch gratings. This is due to the fact that line edges are randomly rough in the mask, which forms a smeared self-image that averages the roughness effectively reducing the . The improvement in the Talbot image is determined by the number of diffraction orders involved which in turn depends on the particular configuration (numerical aperture and periodicity) of the mask [20]. In the simulation, the is improved in self-images for sub 30 nm pitch when EUV radiation is assumed in the illumination.

Acknowledgment

The authors would like to thank E. Neumann at the Helmholtz Nanoelectronic Facility, Forschungszentrum Juelich, for help with FIB milling.

FOOTNOTES

The work of H. Kim was supported in part by the EU FP7 Erasmus Mundus Joint Doctorate Programme EXTATIC under framework partnership agreement FPA-2012-0033 and in part by the COST Action MP1203. The work of L. Juschkin was supported by the Helmholtz Association for a Helmholtz Professorship as a part of the Initiative and Networking Fund. The work of W. Li and M. Marconi was supported by the National Science Foundation through Award ECCS 1507907.

REFERENCES

1. D. Grützmacher "Three dimensional Si/Ge quantum dot crystals" *Nano Lett., vol. 7, no. 10, pp. 3150-3156, 2007*

2. H. F. Talbot "LXXXVI. Facts relating to optical science no. IV" *Philos. Mag., vol. 9, no. 56, pp. 401-407, 1836*

3. L. Rayleigh "XXV. On diffraction-gratings, and on some phenomena connected therewith" *Philos. Mag., vol. 11, no. 67, pp. 196-205, 1881*

4. A. Isoyan "Talbot lithography: Self-imaging of complex structures" *J. Vac. Sci. Technol. B, Microelectron. Nanometer Struct., vol. 27, no. 6, pp. 2931-2937, 2009*

5. H. H. Solak and Y. Ekinci "Achromatic spatial frequency multiplication: A method for production of nanometer-scale periodic structures" *J. Vac. Sci. Technol. B, vol. 23, no. 6, pp. 2705-2710, 2005*

6. L. Urbanski "Defect-tolerant extreme ultraviolet nanoscale printing" *Opt. Lett., vol. 37, no. 17, pp. 3633-3635, 2012*

9

CHAPTER

Nanostructuring of Palladium with Low-Temperature Helium Plasma

P. Fiflis *, M.P. Christenson, N. Connolly and D.N. Ruzic

Department of Nuclear, Plasma and Radiological Engineering, Center for Plasma Material Interactions, University Illinois at Urbana-Champaign, Urbana 61801, IL, USA

Abstract

Impingement of high fluxes of helium ions upon metals at elevated temperatures has given rise to the growth of nanostructured layers on the surface of several metals, such as tungsten and molybdenum. These nanostructured layers grow from the bulk material and have greatly increased surface area over that of a not nanostructured surface. They are also superior to deposited nanostructures due to a lack of worries over adhesion and differences in material properties. Several palladium samples of varying thickness were biased and exposed to a helium helicon plasma. The nanostructures were characterized as a function of the thickness of the palladium layer and of temperature. Bubbles of ~100 nm in diameter appear to be integral to the nanostructuring process. Nanostructured palladium is also shown to have better catalytic activity than not nanostructured palladium.

Keywords: palladium; nanotendrils; helium bubbles; catalysis

1. Introduction

Experiments at several different institutions have observed the growth of tungsten nanostructures under exposure to helium plasmas while investigating the viability of tungsten for high heat flux components in nuclear fusion reactors [1,2]. While these structures are potentially fatal to the fusion plasma when grown [3], they do exhibit characteristics that could be exploited in other applications. A high porosity, a low density of about 10% of the bulk material, large surface area, increased emissivity, and decreased reflectance are all properties of the nanostructured surface [4,5,6,7]. The nanostructures are produced by prolonged exposure to a flux of helium ions while the tungsten is at an elevated temperature. Several studies have investigated the similar formation of nanostructures on metals other than tungsten. Exposure to fluxes of helium in excess of 10^{20} $m^{-2} \cdot s^{-1}$ at a temperature approximately between 30% and 50% of the melting temperature yields several different nanostructures, *i.e.*, cones/pillars have been observed on copper [8], fuzz on molybdenum and tungsten [5], and roughening of the surface on titanium [8]. Palladium is widely used as a catalyst in a variety of different chemical reactions; and because catalytic activity scales linearly with the surface area of the catalyst, experiments were undertaken to investigate the nanostructuring of palladium under irradiation by helium ions. To this end, several palladium substrates were exposed at elevated temperatures to helium ions from a helicon plasma source. The dependence of the formed nanostructures on temperature and sample geometry are described herein.

2. Experimental Section

Palladium samples (Alfa Aesar 99.9%, Ward Hill, MA, USA) were exposed inside of a commercial grade helium helicon source (MORI 200, Trikon Technologies, Newport, UK) [9]). The plasma conditions for the experiments described herein were generated with an RF power of 700 W, a magnetic field of 120 G, and a background helium pressure of 100 mTorr as read by a convectron gauge (Granville Phillips 375, MKS Instruments, Andover, MA, USA). A photo of the experimental chamber can be seen in Figure 1. The resulting plasma density is 1×10^{18} m^{-3} with an electron temperature of 4 eV diagnosed with an RF compensated Langmuir probe [10] in the region where the sample was placed. The palladium was supported via a copper sample holder, which suspended the sample in the plasma. The sample was biased to negative 40 V with respect to plasma potential such that the incoming helium ion flux had an energy of 40 eV and a flux of 2.5×10^{21} $m^{-2} \cdot s^{-1}$. Sample temperature was achieved merely by heating of the sample via the incoming ion flux. Regulation of the temperature, however, was achieved by adjusting the area of the sample in direct contact with the copper sample holder, thereby controlling conduction losses. It should be noted that the centerline density of the plasma is constant to within measurement error over the range of sample placements, so adjustment of the sample relative to the copper sample holder did not change the flux to the target. Temperatures were not directly measured, but rather computed via an experimentally calibrated finite difference model which balances input energy from helium ion irradiation and losses via conduction and radiation [11].

Scanning electron microscopy (Hitachi S4700, Tokyo, Japan) was performed on the exposed samples. Four different geometries were tested; a plate of palladium 1 cm × 2 cm × 0.5 mm, a wire 0.5 mm diameter × 20 cm in length, and two thin films deposited on 25 mm × 25 mm glass substrates with thicknesses of 300 nm (evaporation coating) and 30 nm (magnetron sputter coating).

Figure 1. Photo of exposure chamber showing MORI (Trikon Technologies, Newport, UK) automated matching network, exposure volume, load lock gate valve, and transfer arm for introducing samples without breaking vacuum.

3. Results and Discussion

3.1. Palladium Nanostructuring as a Function of Temperature

Scanning electron microscope (SEM) micrographs of the exposed wire and plate are shown in Figure 2, Figure 3 and Figure 4, respectively. Figure 3 is a series of top down (0° from normal) micrographs of the plate; Figure 4 is a series of micrographs of the plate and has a tilt of 40° from normal introduced to the sample. From these micrographs it can be seen that exposure of palladium at elevated temperatures to fluxes of helium ions forms a series of pillars or tendrils from the surface. Energy dispersive X-ray Spectroscopy (EDX) analysis of these tendrils confirms that they are palladium. These tendrils are 350 ± 100 nm in diameter, 1000 ± 250 nm in height, and possess an areal density of approximately 1.5 tendrils/μm^2. It can also be seen from the difference between the 0° and 40° micrographs of the plate that these tendrils grow normally to the surface. The tendrils are also mostly straight at temperatures less than 700 K, and in excess of 700 K the structures begin to bend and fold. As the temperature increases beyond this (>775 K), the nanostructures begin to increase in diameter (500 ± 100 nm), but not length. The areal density of tendrils decreases to approximately 1 tendril/μm^2. As the temperature exceeds 850 K, the nanostructures appear to begin to anneal back into the bulk, resulting in very rounded and thick tendrils at 880 K (see Figure 2), and full disappearance of tendrils by 900 K (Figure 5). The nanostructuring process appears to be bubble driven, similar to the growth of tungsten nanostructured "fuzz" which is predicted to grow under certain conditions in the divertor region of fusion reactors [12]. However, while the bubbles that drive nanostructuring in tungsten are approximately 10 nm in diameter, the pits observed in the surface of the nanostructured palladium (attributed to bubble bursting at the surface similar to tungsten [11]) are 75 ± 25 nm in diameter at temperatures less than 750 K. At temperatures above 750 K, these pits swell in size to 110 ± 30 nm in diameter. Normalizing two characteristic parameters of the nanostructures to the pit diameter draws striking parallels between tungsten and palladium nanostructuring. The ratio of tendril diameter to pit diameter in tungsten is approximately 3 to 4 [11]. Similarly, the ratio of tendril diameter to pit diameter in palladium is also 3 to 4. Additionally, the separation distance between individual tendrils is approximately 7 to 11 times the diameter of the pits in tungsten [11]. With a tendril separation distance of 800 ± 150 nm, palladium nanostructures have a ratio of tendril separation distance to pit diameter in the exact same range. This is very indicative of a similar bubble mechanism driving the growth of fuzz tendrils, whereby bubbles created in the bulk of the material effectively rise to the surface or grow and thin the material above them, subsequently rupturing. As more bubbles impact the surface, hills and valleys start to form stochastically. Bubbles then are more likely to connect to a valley rather than a hill by virtue of shorter path length, and as a result nanostructures grow. Characteristics such as the ratio of tendril diameter or tendril separation distance to the pit diameter fall out of a simple Monte Carlo model implementing only the assumption that bubbles are more likely to rupture at a valley than a hill [11]. Since the model is independent of material properties, any nanostructuring via the same mechanism will show similar tendril to pit diameter ratios and tendril separation to pit diameter ratios, as is seen here. Much like tungsten nanostructuring via helium plasma bombardment, palladium nanostructuring appears to have a window of temperature for which it can grow tendrils. Tungsten fuzz grows within the temperature range of 1000–2000 K (0.27–0.54 T_m) [13]. Palladium nanostructuring in the experiments described herein occurred at 650 K (0.33 T_m) and 880 K (0.48 T_m) as well as several other intermediate temperatures and was bracketed by a lack of nanostructuring formation at 500 K (0.27 T_m) and 900 K (0.49 T_m). When normalized to the melting point of the material, the active temperature range for the bubble driven nanostructuring appears to be very similar.

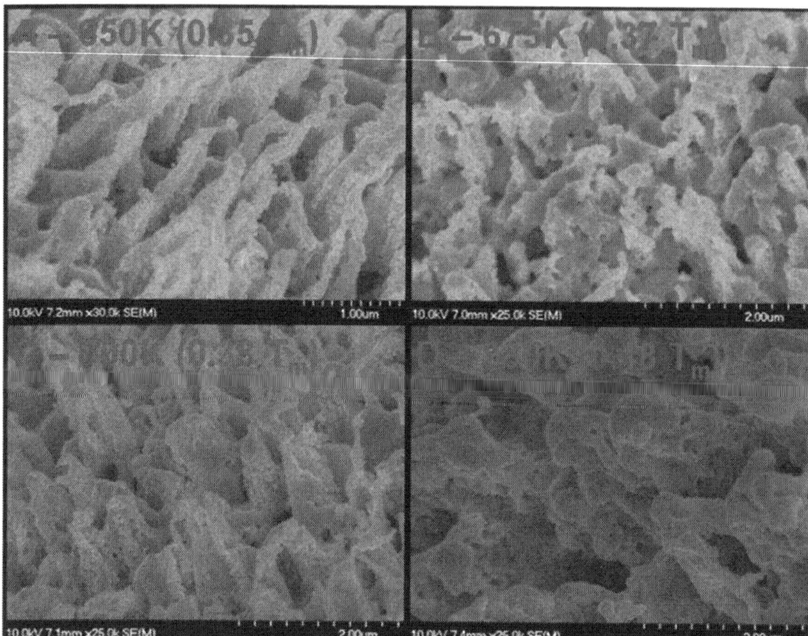

Figure 2. Scanning electron microscope (SEM) micrographs of palladium surface (0.5 mm diameter wire sample) after exposure to helium plasma at elevated temperature. The flux to each area is identical, the only changed variable is temperature (noted in the upper left corner of each micrograph both absolute and as a fraction of the melting point of palladium).

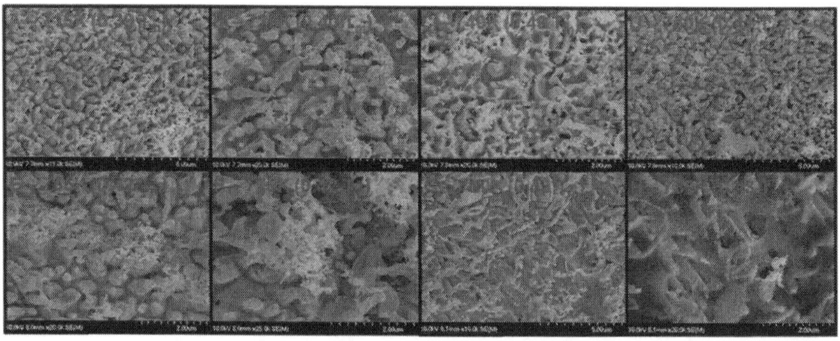

Figure 3. SEM micrographs of palladium surface (0.5 mm plate sample) after exposure to helium plasma at elevated temperature. The flux to each area is identical, the only changed variable is temperature (noted in the upper left corner of each micrograph both absolute and as a fraction of the melting point of palladium). Secondary electron collection performed at a tilt angle of 0° with respect to the surface normal.

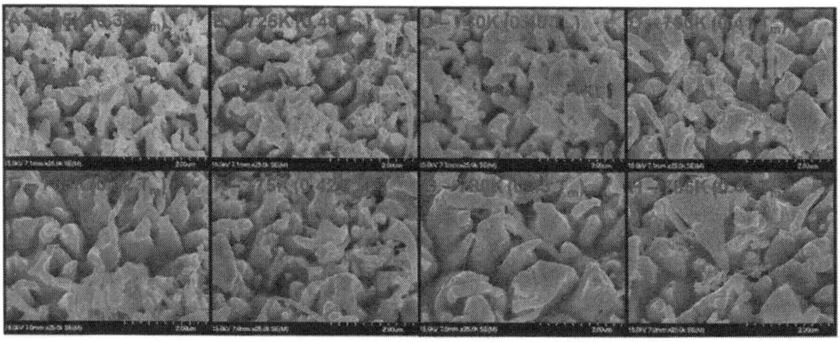

Figure 4. SEM micrographs of palladium surface (0.5 mm plate sample) after exposure to helium plasma at elevated temperature. The flux to each area is identical, the only changed variable is temperature (noted in the upper left corner of each micrograph both absolute and as a fraction of the melting point of palladium). Secondary electron collection performed at a tilt angle of 40° with respect to the surface normal.

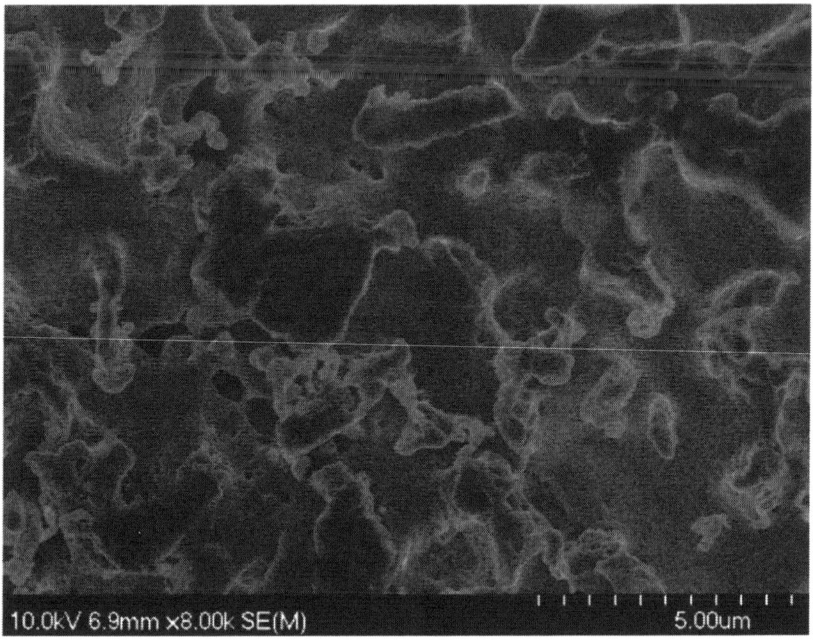

Figure 5. SEM micrograph of palladium surface (0.5 mm diameter wire sample) after exposure to helium plasma at 900 K, only a couple tendrils are visible as the annealing rate of the tendrils begins to exceed the rate of growth.

3.2. Palladium Nanostructuring versus Palladium Thickness

Many applications of palladium nanostructuring revolve around the increased surface area to volume ratio (*i.e.*, improving the catalytic activity for a given weight of palladium). As a result of this, the ability to grow nanostructures on thin films of palladium is very desirable as this would further increase the surface area to volume ratio of an amount of palladium with an already large surface area to volume ratio. There, however, was thought to be a minimum thickness at which the palladium would no longer nanostructure because the bubbles necessary to drive the growth of palladium nanostructures were approximately 100 nm in diameter and could therefore not grow to full size in very thin films. As thick substrates of 0.5 mm thickness and diameter had already been tested, two thin films (of thickness 300 and 30 nm) were investigated to see the structures that would form. SEM micrographs of each sample can be seen in Figure 6 and Figure 7, respectively. Nanostructure growth for the 300 nm film appears to be again bubble driven. Tendrils of diameter 320 ± 100 nm are apparent and so are pits of 75 ± 15 nm diameter. These are commensurate with the diameters of the tendrils and pits of the bulk samples at the same temperature of exposure. The tendrils are also very straight, much like the tendrils observed in the bulk samples at temperatures <700 K. This implies the bubble formation depths and subsequent loop punching to the surface occurs at depths of less than 3 bubble diameters, much like the formation of helium bubbles within tungsten which drive nanostructuring [11]. As the thickness of the film is reduced below that of the bubble diameter, formation of full helium bubbles to drive nanostructuring is suppressed. Instead, it appears as though formation and growth of bubbles within the 30 nm thick film rupture the film without being able to build upon each other and grow nanostructures. This results in a series of pits in the surface, but no vertical growth of nanostructures. These pits are of diameter 130 ± 35 nm, which is larger than those observed in the 300 nm and bulk samples. Wrinkles are also evident in the palladium film which is indicative of delamination of the palladium film from the SiO_2 substrate. The palladium film was deposited via magnetron sputtering at room temperature and due to residual tensile stresses in the film, once it became delaminated, it wrinkled.

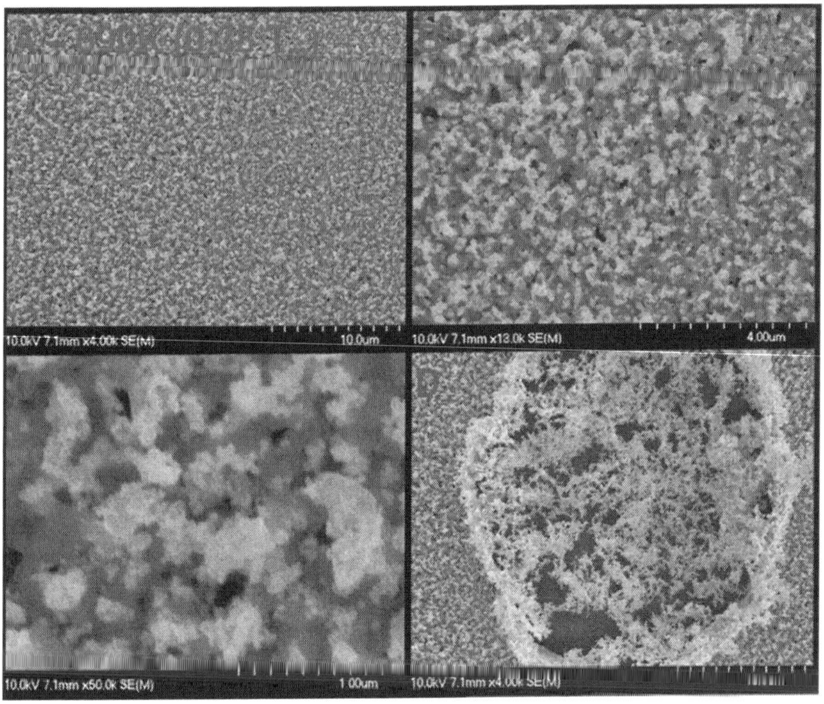

Figure 6. SEM micrograph of palladium surface (300 nm thin film deposited on SiO_2) after exposure to helium plasma at elevated temperature. (**A**), (**B**), and (**C**) are different resolutions of the same location showing growth of tendrils and voids that appear to penetrate down to the SiO_2 substrate. Tendrils approximately the same diameter as those observed on bulk Pd samples are observed. Pits of similar diameter are also observed. (**D**) shows an area of the palladium film where the helium plasma has eroded through the palladium film to the substrate with very thin tendrils of Pd stretching across.

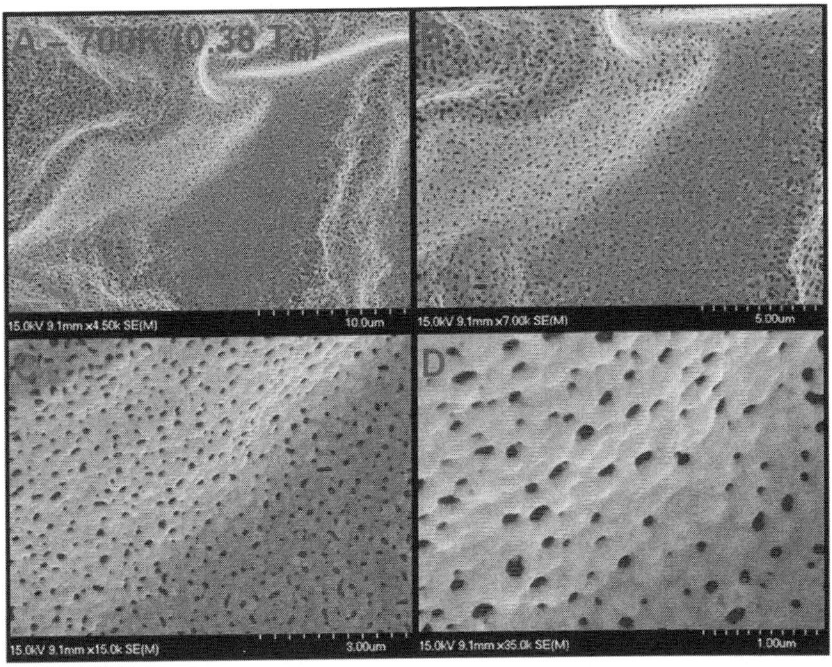

Figure 7. SEM micrograph of palladium surface (30 nm thin film deposited on SiO_2) after exposure to helium plasma at elevated temperature. Figure 5A–D are different resolutions of the same location showing growth no tendril growth, but a significant amount of voids. These voids are of a diameter greater than the pits observed in the bulk and 300 nm film samples. Large wrinkles appear evident in the film. It appears as though formation and growth of bubbles within the 30 nm thick film rupture the film without being able to build upon each other and grow nanostructures.

3.3. Catalysis with Nanostructured Palladium

It has been suggested that the increased surface area of nanostructured palladium could provide greatly enhanced catalytic activity. To investigate this hypothesis, a comparison was made between identical plates, one not nanostructured and the other nanostructured under identical conditions to the plate described above. This comparison of catalytic properties was carried out using a reduction reaction, which modified cyclohexene into cyclohexane through the syn addition of two hydrogens using the palladium pieces as the catalysts for the reaction [14]:

$$C_6H_{10} + H_2 \rightarrow C_6H_{12}$$

The catalytic properties of each of the palladium samples, both smooth and nanostructured, were compared to the industrial palladium catalytic standard of palladium absorbed onto a carbon surface [15]. This standard is widely accepted as the best way to increase surface area for catalytic reactivity.

Alkenes are reduced to alkanes through a multi-body process, where the palladium or other catalyst acts as an activation site for the reaction. Following the Horiuti-Polanyi mechanism [14], hydrogen first dissociatively chemisorbs to the bonding site on the palladium surface, but only if the hydrogen molecule has its axis parallel to the surface of the palladium crystal [16]. For this reason, hydrogen is often found in excess in these reactions. Cyclohexene, present in the reaction in liquid form, dissociatively bonds to the surface of the palladium in much the same way as the hydrogen molecule, where the pi bond in the double bond between the carbons is broken. The carbons then bond to the palladium surface. The dissociated hydrogen atoms then bond to the free sites on the carbon atoms currently bonded to the palladium surface and the molecule leaves the palladium surface as an alkane since alkanes are not strongly adsorbed on the surface. For a single reaction on a simple palladium surface, this leads to a significant reduction in the activation energy, making the reaction feasible for scale-up.

All reactions were carried out with the primary reactant, cyclohexene, in liquid phase. A block diagram of the experimental setup can be seen in Figure 8. The reaction vessel was a 500 mL, three-neck round bottom flask connected to a vacuum line on one neck, a gas bubbler on the second neck, and stoppered on the third neck. Prior to closing off the flask, the catalytic samples were weighed and added, along with a magnetic stir rod. The reaction chamber was then sealed. The chamber was then purged with argon and evacuated three separate times to ensure atmospheric purity. 150 mL of cyclohexene was added through the stopper using a syringe. A second syringe connected to a hydrogen line was then inserted into the cyclohexene liquid in order to bubble into the flask an excess of hydrogen gas, as noted by the bubbler. Before hydrogen was added, a control sample was taken. Samples were then taken using a microsyringe at 30 min intervals for 180 min, while the hydrogen was bubbled through the reactants and the stir rod agitated the reactants. This was done to test the reaction rate for each of the catalyst types. All samples taken were 100 μL in volume and were added to NMR sample tubes along with 600 μL of deuterated chloroform ($CDCl_3$), which were mixed through inversion.

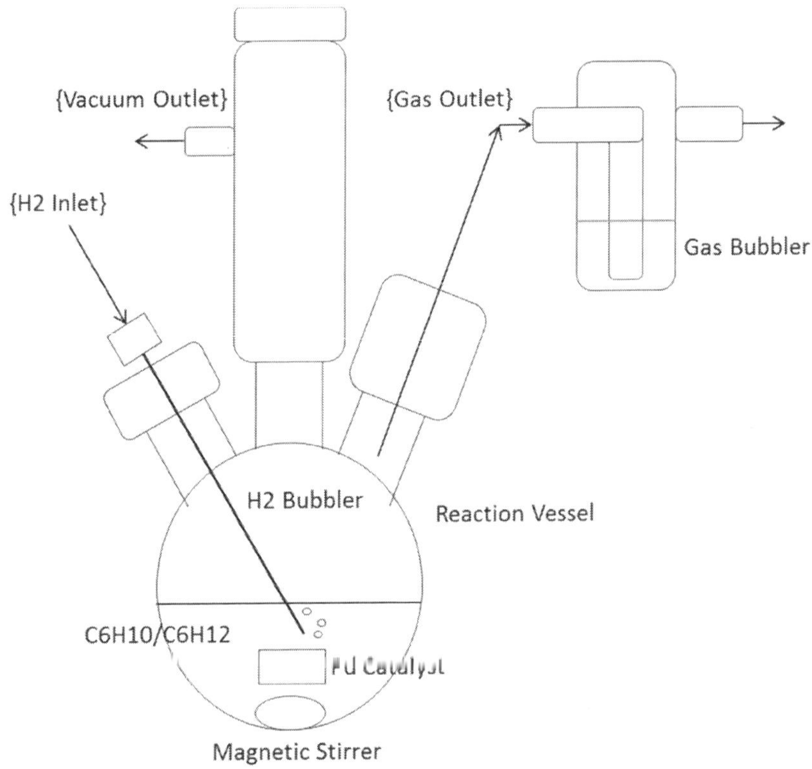

Figure 8. A block diagram of the palladium-catalyzed hydrogenation reaction vessel, including the hydrogen gas inlet, the vacuum cylinder outlet, the magnetic stirrer, and the gas bubbler used to qualitatively determine the flow rate of the hydrogen through the reactant volume.

The catalytic capability of the different palladium types was measured by determining the ratios of the areas under the typical cyclohexene peaks to the typical cyclohexane peaks seen in Nuclear magnetic resonance (NMR) spectra. A Varian Unity Inova NMR spectroscopy machine [17,18] was used to make the measurements, and each scan used a spectral frequency of 399.74 MHz, with 1 transient at an acquisition time of 4.096 seconds. The scans were performed using one-dimensional proton NMR at a 45 degree pulse width. Ratios were taken between areas under the 1.4 ppm cyclohexane peak and the mean of the areas under the 1.2 ppm, 1.7 ppm, and 2.0 ppm cyclohexene peaks, normalized to the deuterated chloroform standard. Figure 9 shows the results of the catalytic conversion over time for the different catalyst types, indicating that the fuzzed sample is greatly improved in conversion over the non-fuzzed sample.

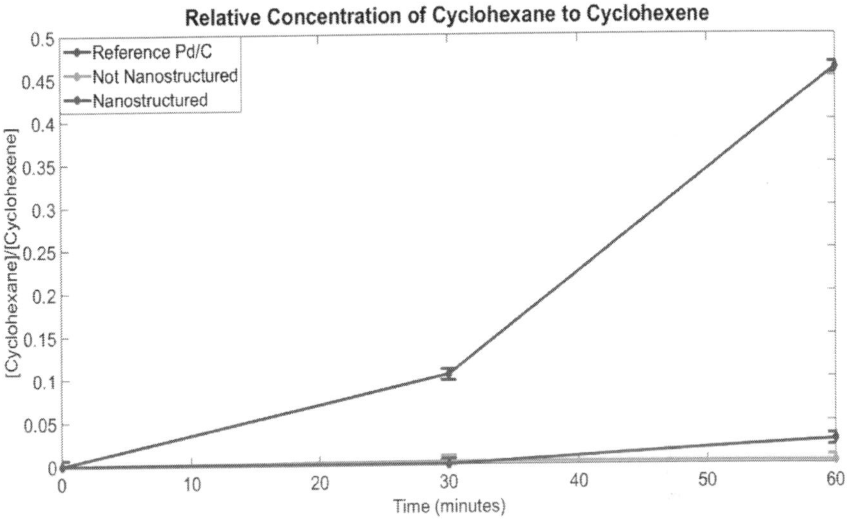

Figure 9. A plot of the yield measured as the ratio of the concentration of the cylcohexane to cyclohexene. This ratio is based on the ratios of areas under the curve of the cyclohexane 1.4 ppm peak and the mean of the intensities of the three cyclohexene peaks seen at 1.2 ppm, 1.7 ppm, and 2.0 ppm in the NMR scans, which are characteristic for the respective compounds and are normalized to the deuterated chloroform standard. These ratios were also taken as a function of time to, not only observe the effectiveness of each catalyst type, but also how the kinetics of the reaction compare with each catalyst type.

4. Discussion and Conclusions

A variety of palladium samples were exposed to a flux of 40 eV helium ions at temperatures between 0.3 and 0.5 T_m. Samples of bulk palladium (*i.e.*, wire and plate) showed evidence of bubbles of approximate diameter 100 nm and tendrils of approximate diameter 350 nm. The nanostructuring growth mechanism appears to be similar to that of tungsten, with an active temperature range similar to that of tungsten after normalization to the melting point. However, the diameter of the bubbles is much larger than that of those observed in tungsten. Previous studies of exposure of different metals to energetic helium fluxes at elevated temperatures have suggested that the nanostructuring process is heavily dependent on crystal structure [8]. Body centered cubic (bcc) crystals, such as tungsten, molybdenum, and tantalum, show very similar nanostructures in both size and morphology. Palladium is a face centered cubic (fcc) material, and therefore, will nanostructure differently than the bcc tungsten. However, since the ratio of tendril diameter to pit diameter as well as the ratio of inter-tendril separation to pit diameter is the same for both tungsten and palladium, it is highly probable that the mechanism for the formation of the nanostructures is the same. The difference in bubble size then is the biggest driver in the difference in observed morphology.

Nanostructuring the palladium plate resulted in a large increase in catalytic activity beyond that of the non-nanostructured sample. It should, however, be noted that the reference Pd/C catalyst outperformed both of the fuzzed and non-fuzzed samples due to the very large surface area provided by the activated carbon. Mechanical removal of these nanostructures from the surface of the palladium to produce a finely nanostructured powder catalyst may increase the catalytic activity beyond that of the standard. Alternatively, nanostructuring of other geometries with a helium plasma may offer advantages in systems where filtration of powder catalysts would be impractical and conventional methods of surface roughening, such as sand blasting, would be too violent on fragile catalyst geometries. Nanostructuring by helium plasma could also be performed atop a sand blasted layer to further increase the surface area.

Acknowledgments

The authors would like to Samuel Yick (CSIRO, Australia) for preparation of the 30 nm thin film palladium sample, as well as Lynford Goddard and Steven McKeown for preparation of the 300 nm palladium sample, and Dean Olson for running NMR tests and providing guidance in NMR analyzation.

Author Contributions

The nanostructuring experiments were conceived and designed by P. Fiflis and D.N. Ruzic. P. Fiflis carried out the nanostructuring experiments. Chemical analysis was designed and performed by M.P. Christenson and N. Connolly. Data analysis performed by P. Fiflis, M.P. Christenson, and N. Connolly. All authors contributed to the composition of the paper.

Conflicts of Interest

The authors declare no conflict of interest.

References

1. Wright, G.M.; Brunner, D.; Baldwin, M.J.; Bystrov, K.; Doerner, R.P.; Labombard, B.; Lipschultz, B.; De Temmerman, G.; Terry, J.L.; Whyte, D.G.; *et al.* Comparison of tungsten nano-tendrils grown in Alcator C-Mod and linear plasma devices. *J. Nuc. Mat.* **2013**, *438*, S84–S89.

2. Baldwin, M.J.; Doerner, R.P. Formation of helium induced nanostructure "fuzz" on various tungsten grains. *J. Nuc. Mat.* **2010**, *404*, 165–173.

3. Pitts, R.A.; Carpentier, S.; Escourbiac, F.; Hirai, T.; Komarov, V.; Lisgo, S.; Kukushkin, A.S.; Loarte, A.; Merola, M.; Sashala Naik, A.; *et al.* A full tungsten divertor for ITER: Physics issues and design status. *J. Nuc. Mat.* **2013**, *438*, S48–S56.

4. Takamura, S.; Ohno, N.; Nishijima, D.; Kajita, S. Formation of nanostructured tungsten with arborescent shape due to helium plasma irradiation. *Plas. Fus. Res.* **2006**, *1*.

5. De Temmerman, G.; Bystrov, K.; Zielinski, J.J.; Balden, M.; Matern, G.; Arnas, C.; Marot, L. Nanostructuring of molybdenum and tungsten surfaces by low-energy helium ions. *J. Vac. Sci. Technol. A* **2012**, *30*, 041306.

6. Baldwin, M.J.; Doerner, R.P. Helium induced nanoscopic morphology on tungsten under fusion relevant plasma conditions. *Nucl. Fus.* **2008**, *48*, 3.

7. Kajita, S.; Ono, N.; Yokochi, T.; Yoshida, N.; Yoshihara, R.; Takamura, S.; Hatae, T. Optical properties of

nanostructured tungsten in near infrared range. *Plas. Phys. Contr. Fus.* **2012**, *54*, 10.

8. Tanyeli, I.; Marot, L.; Mathys, D.; van de Sanden, M.C.M.; De Temmerman, G. Surface modifications induced by high fluxes of low energy helium ions. *Sci. Reports.* **2015**, *5*.

9. Tobe, R.; Sekiguchi, A.; Sasaki, M.; Okada, O.; Hosokawa, N. Plasma-enhanced CVD of TiN and Ti using low-pressure and high-density helicon plasma. *Thin Solid Films* **1996**, *281–282*, 155–158.

10. Ruzic, D.N. *Electric Probes for Low Temperature Plasmas*, 1st ed.; American Vacuum Society Education Committee: New York, NY, USA, 1994.

11. Fiflis, P.; Curreli, D.; Ruzic, D.N. Direct Time-Resolved Observation of Tungsten Nanostructured Growth Due to Helium Plasma Exposure. *Nucl. Fus.* **2015**, *55*, 3.

12. Brooks, J.N.; Allain, J.P.; Doerner, R.P.; Hassanein, A.; Nygren, R.; Rognlien, T.D.; Whyte, D.G. Plasma–surface interaction issues of an all-metal ITER. *Nucl. Fus.* **2009**, *49*, 035007.

13. Ueda, Y.; Peng, H.Y.; Lee, H.T.; Ohno, N.; Kajita, S.; Yoshida, N.; Doerner, R.; De Temmerman, G.; Alimov, V.; Wright, G. Helium effects on tungsten surface morphology and deuterium retention. *J. Nuc. Matl.* **2013**, *442*, S267–S272.

14. Horiuti, I.; Polanyi, M. Exchange Reactions of Hydrogen on Metallic Catalysts. *T. Faraday Soc.***1934**, *30*, 1164–1172.

15. Puskás, R.; Sápi, A.; Kukovecz, A.; Kónya, Z. Comparison of Nanoscaled Palladium Catalysts Supported on Various Carbon Allotropes. *Top. Catal.* **2012**, *55*, 865–872.

16. Lischka, M.; Groβ, A. Hydrogen on palladium: A model system for the interaction of atoms and molecules with metal surfaces. *Recent Dev. Vac. Sci. Technol.* **2003**, *37*, 111–132.

17. McKenna, S.; Spyracopoulos, L.; Moraes, T.; Pastushok, L.; Ptak, C.; Xiao, W.; Ellison, M.J. Noncovalent Interaction between Ubiquitin and the Human DNA Repair Protein Mms2 Is Required for Ubc13-mediated Polyubiquitination. *J. Bio. Chem.* **2001**, *276*, 40120–40126.

18. Miura, K.; Ohgiya, S.; Hoshino, T.; Nemoto, N.; Suetake, T.; Miura, A.; Spracopoulos, L.; Kondo, H.; Tsuda, S. NMR Analysis of Type III Antifreeze Protein Intramolecular Dimer. *J. Bio. Chem.***2001**, *276*, 1304–1310.

Index